SpringerBriefs in Applied Sciences and Technology

Thermal Engineering and Applied Science

Series editor

Francis A. Kulacki, Minneapolis, USA

For further volumes:
http://www.springer.com/series/10305

Junjie Gu · Zhongxue Gan

Entransy in Phase-Change Systems

 Springer

Junjie Gu
Department of Mechanical
   and Aerospace Engineering
Carleton University
Ottawa, ON
Canada

Zhongxue Gan
ENN Intelligent Energy Co., Ltd.
ENN Group
Langfang
China

ISSN 2193-2530   ISSN 2193-2549   (electronic)
ISBN 978-3-319-07427-6   ISBN 978-3-319-07428-3   (eBook)
DOI 10.1007/978-3-319-07428-3
Springer Cham Heidelberg New York Dordrecht London

Library of Congress Control Number: 2014941102

© The Author(s) 2014
This work is subject to copyright. All rights are reserved by the Publisher, whether the whole or part of the material is concerned, specifically the rights of translation, reprinting, reuse of illustrations, recitation, broadcasting, reproduction on microfilms or in any other physical way, and transmission or information storage and retrieval, electronic adaptation, computer software, or by similar or dissimilar methodology now known or hereafter developed. Exempted from this legal reservation are brief excerpts in connection with reviews or scholarly analysis or material supplied specifically for the purpose of being entered and executed on a computer system, for exclusive use by the purchaser of the work. Duplication of this publication or parts thereof is permitted only under the provisions of the Copyright Law of the Publisher's location, in its current version, and permission for use must always be obtained from Springer. Permissions for use may be obtained through RightsLink at the Copyright Clearance Center. Violations are liable to prosecution under the respective Copyright Law. The use of general descriptive names, registered names, trademarks, service marks, etc. in this publication does not imply, even in the absence of a specific statement, that such names are exempt from the relevant protective laws and regulations and therefore free for general use.
While the advice and information in this book are believed to be true and accurate at the date of publication, neither the authors nor the editors nor the publisher can accept any legal responsibility for any errors or omissions that may be made. The publisher makes no warranty, express or implied, with respect to the material contained herein.

Printed on acid-free paper

Springer is part of Springer Science+Business Media (www.springer.com)

Preface

This monograph presents recent developments in our research on a new thermo-dynamic concept—Entransy, specifically for phase change processes, such as boiling, condensation, absorption and desorption, etc. This body of work can be used by university postgraduate students as a textbook, by students for self-study, by researchers and professors as an academic reference book, and by engineers and designers as a guide for efficient energy system development.

The primary goal is to summarize our research results in energy systems where phase changes often occur. Entransy is a useful concept/tool for optimal design of such energy systems.

The book includes four chapters: (1) Introduction of the concept—entransy; (2) Fundamentals of Entransy and Entransy Dissipation Theory; (3) Application of Entransy Theory in Thermal Storage System; and (4) Application of Entransy Theory in Absorption Refrigeration System.

Here, we would like to express our sincere appreciation to Dr. Linghui Zhu, who offered great help in the scientific contribution and the careful editing work, which is well reflected in this book. We would like to thank Mr. Hooman Abdi for his enthusiasm and scientific contributions during his thesis work at Carleton University, Canada, which formed important parts of this work.

We also would like to express our appreciation for the inspirational environment provided by ENN Group, China, where some of this work was completed.

Finally, we would like to express our appreciation to our wives, Jane and Xiaoshu, and to our children for their continued patience, understanding, and support throughout the preparation of this book.

Junjie Gu
Zhongxue Gan

Acknowledgments

Contributions from Dr. Linghui Zhu and Mr. Hooman Abdi are highly appreciated and acknowledged.

Contents

Nomenclature

A   Area (m^2)

c   Specific heat (J/kgK)

d   Capsule diameter (m)

D   Tank diameter (m)

e   Energy (J)

Fc   Field synergy number

g   Acceleration of gravity (m/s^2)

G   Entransy (WK)

h   Convective heat transfer coefficient (W/m^2 K), enthalpy (J)

J   Bessel function

k   Thermal conductivity (W/m.K)

L   Latent heat of melting (J/kg)

Nu   Nusselt number

Pr   Prandtl number

Q   Heat transfer (J)

q   Heat flux (J/m^2)

r   Radius, radial distance (m)

R   Thermal resistance (K/W), tank radius (m)

Re   Reynolds number

S   Source term

T   HTF temperature (°C)

t   Time (s)

U   Velocity (m/s)

u   Velocity (m/s)

V   Volume (m^3)

W   Work (J)

x   X-axis direction, longitudinal distance (m)

Y   Y-axis direction

Greek Symbols

ε	Porosity
η	Thermal efficiency
μ	Dynamic viscosity
π	Pi number
θ	PCM temperature
ρ	Density
Ξ	Exergy (J)
ψ	Exergy efficiency

Subscripts

$*$	Normalized value
,	Partial derivative
_	Average value
2	Two-phase
cap	Capsule
CV	Control volume
e	Exit
hom	Homogeneous
i	Internal, node number, inlet port
ice	Ice
inl	Inlet
int	Initial
l	Liquid
max	Maximum
n	Normal component, general direction
nucl	Nucleation
out	Outlet
p	PCM, particle
p,l	Liquid PCM
p,s	Solid PCM
r	Respect to r
s	Solid
sup	Supercooling
tot	Total
w	Water
x	Respect to x

Superscripts

+	Normalized value

Acronyms

A/C	Air conditioning
CTES	Cold thermal energy storage
HTF	Heat transfer fluid
HVAC	Heating, ventilation and air conditioning
LHTES	Latent heat thermal storage
NTU	Number of transfer unit
PMC	Phase change material
TES	Thermal energy storage

Chapter 1
Introduction

Abstract A thermal energy storage system can be regarded as a heat exchanger device which has the heat transfer between the heat transfer fluid and the phase change materials. In this chapter, the fundamental for energy storage system, as well as the modeling method for thermal system is introduced. The most recent entransy analysis methodology is described and compared with previous energy and exergy theories, and a detailed literature review is presented on entransy dissipation theory.

Keywords Thermal storage system · Thermal energy storage · Entransy · Exergy · Exergy analysis · Heat transfer · Entransy dissipation

Energy crisis, global warming, ozone layer depletion, and many other crises attract the attention of researchers to find new ways for a sustainable development of our society. During the past decades, well-defined energy and exergy analyses have been used to assess the performance of many thermal systems. Advantages of these types of methods have been discussed in substantial research works. Exergy analysis and consequently entropy generation are two main tools extensively used for thermal assessments, especially for the engineering systems dealing with heat–work interactions such as in engines and turbines. In this work, new parameters and assessment tools recently introduced and used for thermal system performances, e.g., entransy, etc., have been implemented for numerical experiments and assessments.

Any thermal energy storage (TES) system can be regarded as a heat exchanger device which has the heat transfer between the heat transfer fluid (HTF) and the phase change materials (PCM) (usually through a capsule skin) as the main mechanism encountered to define the whole process. A better understanding of the phenomena and better assessment tools are vital. There are many numerical and experimental studies published in TES system applications. Here is a short review of the most recent thermal assessment tools followed by the studies of numerical experiments with the new assessment tools for heat transfer, heat exchangers, and thermal energy storages.

J. Gu and Z. Gan, *Entransy in Phase-Change Systems*,
SpringerBriefs in Thermal Engineering and Applied Science,
DOI: 10.1007/978-3-319-07428-3_1, © The Author(s) 2014

Guo et al. introduced the new heat transfer parameter named as "entransy," which is equivalent to potential energy in electricity [1]. Entransy, G, is equal to $\frac{1}{2}QT$ where Q represents an equivalent of electrical charge equal to mcT in heat transfer applications. The new quantity is equivalent to the electrical energy stored in a capacitor in electricity. The entransy term itself is selected in a way that shows the energy (en-) and transfer ability (-transy) of an object or system. The potential ability for heat transfer of an object with higher value of entransy is greater than that of a lower entransy object, like two electrical capacitors at two different potential energy levels. In order to maintain [2] a heat transfer flow, entransy should be consumed and it would indicate the level of heat transfer irreversibility. It opens new areas for the heat transfer assessment and optimization. The extremum principle of entransy dissipation has been developed, and novel configurations have been proposed for heat transfer enhancement in conduction and convection modes [1].

Guo et al. [3] defined equivalent thermal resistance based on entransy dissipation and used it in heat exchanger design concept. It is a measure of what it is needed to pay to maintain the heat flow rate as desired. Thermal resistance of the heat exchanger has been related to the thermal effectiveness of the heat exchanger which is a measure of maximum possible heat transfer between the cooling or heating surface and the heat transfer fluid. Based on the new quantity and level of irreversibility in heat transfer taking place between the hot and cold fluids in the heat exchanger, there is no entropy generation paradox trend appears. It is suggested to analyze these types of thermal systems based on thermal entransy dissipation rather than entropy generation definition. In another study, Chen et al. [4] analyzed a heat exchanger based on entransy dissipation and its equivalent thermal resistance analytically and numerically. It is also shown that minimum entropy generation rate does not always correspond to a high rate of heat transfer and entransy dissipation extremum is a better tool to assess the heat transfer phenomena in heat exchangers.

Chen et al. compare entropy- and entransy-based optimization principles for convective heat transfer applications. In thermal engineering systems which have no work–heat interactions, the entransy optimization method has been found quite suitable [5]. A two-dimensional convective heat transfer case has been also analyzed through these methods to optimize the heat transfer numerically.

Jia et al. minimized the power consumption of a convective heat transfer flow based on an optimization procedure in which the power consumption has been considered as the objective function and the entransy consumption as constraint [6]. Nusselt number can be doubled by proper positioning of vortexes in the flow field but at the same time, the flow resistance (power consumption) increased by 25 %. This is a numerical verification on how to improve heat transfer without significant additional payment and costs (pumping power increment).

Wei et al. used an optimization procedure based on minimum entransy dissipation in a tubular heat exchanger filled with porous medium with and without highly conductive fins [7]. Heat transfer will be enhanced in the presence of the fins, and the ratio of the inner to the outer radius of the tube has been optimized in these two cases. The analytical solutions have been found for pure conduction heat transfer mode and a constructal design procedure has been followed.

Liu et al. developed entransy transport formulation by multiplying both sides of the energy equation by T and also introduced entransy of incompressible fluids $dz(= T\,dh = cT\,dT)$ as a state variable [8], which is similar to the entropy definition in thermodynamics. Optimum cavity flow velocity fields have been also found for different boundary conditions based on the Lagrange conditional extremum principle with the help of minimum entransy principle. It is also showed that optimization based on entransy dissipation is more useful than entropy minimization principle for these types of thermal systems. Wei et al. used the entransy dissipation and the equivalent thermal resistance to optimize volume to point constructal problems [9]. Wu and Liang used the same entransy dissipation principle to optimize heat transfer in radiation mode. Entransy dissipation minimization method has been used for all three modes of heat transfer [10].

Tao et al. extended the use of field synergy to more applicable heat transfer elliptic problems [11]. The idea has been initiated from analogy between boundary layer (parabolic convection) and conduction heat transfer modes (see also [12]). The synergy between velocity and temperature gradient fields is one of the assessment tools to measure the strength of heat transfer, especially for high Peclet number (>100) flows. Guo et al. [13] introduced a non-dimensional field synergy number (Fc) and showed that for the best convective heat transfer as an ideal case, this number should be equal to 1. The non-dimensional area integration of $U \cdot \nabla T$ within a boundary layer on a heating or cooling surface should be equal to Nu/RePr and for an ideal (maximum) heat transfer case the value should be equal to one. In many practical applications, the Fc number is far from unity and it shows a substantial potential for heat transfer improvement. Figure 1.1 shows the Fc ranges for different flow regimes at different Re numbers. The concept of synergy between the two aforementioned fields (fields' coordination) opens new understanding to the convective heat transfer and provides researchers with a measurement tool for the improvement in heat transfer assessments. A couple of numerical experiments have been also considered, and heat transfer performance of a new tube design has been investigated and compared with smooth tube.

Guo et al. did a thorough optimization on a shell-and-tube heat exchanger with the help of field synergy number (Fc) [14]. The genetic algorithm has been used for the optimization on the baffled heat exchanger and the method is found to be better than traditional optimizations where total cost has been considered as the objective function. The optimal design offers lower cost in the case of maximum field synergy number.

During the past decades, many types of energy storage materials and techniques have been investigated and introduced to the industries to provide a relationship of daily energy consumption by the refrigeration or air-conditioning (A/C) units. In order to analyze the storage system performance, a proper modeling tool based on the fluid flow and heat transfer governing equations would be needed. Then, a thermodynamic analysis according to the energy and exergy assessments is usually

Fig. 1.1 Variation of Fc
number in different
convection mode of
engineering applications: *1*
fully coordinated flow or
ideal convection, *2* laminar
boundary layer, *3* turbulent
boundary layer, and *4*
turbulent convective heat
transfer in circular tube [13]

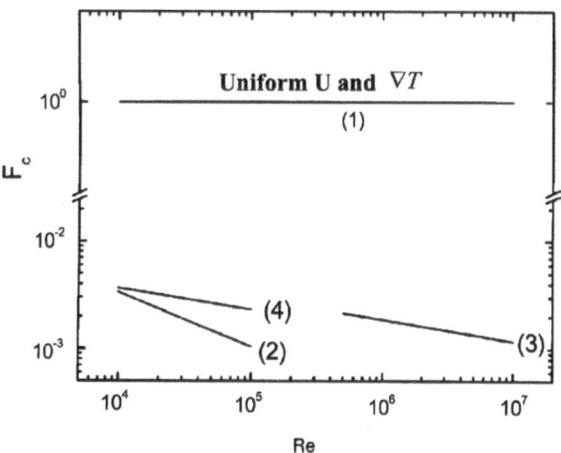

performed to find out how efficient the storing is a one-dimensional energy equation has been solved for a thermal energy storage system containing paraffin spherical capsules. The effect of inlet heat transfer fluid temperature and its mass flow rate, capsule diameter, and fixed and variable phase change temperature of the PCM on the overall thermal performance have been investigated for charging and discharging modes [15]. Charging and discharging times have been found lower in experiments with smaller capsule diameters. The temperature difference of the PCM and inlet HTF (higher Stefan number) also decreases with the frequency of charging and discharging periods. The melting time has observed to be lower in case of using a paraffin with variable melting temperature in comparison with the cases with a constant melting temperature material.

Amri et al. investigated thermal performance of a TES with PCM in horizontal rectangular slab capsules [16]. A couple of parameters such as PCM thickness, vertical distance between the slabs and the mass flow rate have been considered for optimization. A numerical model has been used to solve the energy equation validated with experimental results. They showed that with a proper optimized design, it is possible to achieve a storage density as high as 53–83 % of the PCM latent capacity.

A shell-and-tube heat exchanger consisting a phase change material on the outer side has been experimentally analyzed by Akif et al. [17] in order to assess the performance of the system thermodynamically and experimentally. Different metals have been used for the wall of the heat exchanger and the HTF enters into the system at different temperatures with different mass flow rates. In charging mode, energetic efficiency decreases with higher HTF inlet temperatures, higher flow rates, and higher shell diameters. Higher exergetic efficiency can be obtained in cases with lower temperature difference between the HTF inlet and the PCM temperatures.

In this chapter, newly defined variables are implemented for thermal assessment and three-dimensional numerical models are used to assess thermal performance of the thermal systems.

1.1 Fluid Flow and Heat Transfer Modeling

Modeling of an engineering system is based on the application of the basic conservation laws as the first important step of analytical or numerical study. It needs a bounded region as a control volume. A simple presentation of a control volume and its interaction with the surrounding is shown schematically in Fig. 1.2. It can be a small or infinitesimal portion of a system or even a typical thermal engineering system (e.g., different types of pumps, heat exchangers, or cold thermal energy storage (CTES) tanks). For example, in finite control volume method in computational fluid dynamics (CFD), the governing (mass, momentum, and energy) equations have been discretized over a gridded region consisting a number of control volumes (CV) represented by nodes having a mean and constant intensive properties such as temperature and pressure according to the mass and energy interactions of each node (CV) with their surrounding (as boundary conditions) and neighboring nodes.

1.2 Geometry and Performance Optimization

Geometry and configuration of any thermal system such as CTES systems are of the main factors on the overall performance. Bejan et al. [18] discussed in details about the main parameters confining an optimum shape in an engineering or natural system by heat transfer, fluid flow, and thermodynamic principles, connecting a shape configuration optimization process to an engineering problem based on energy and exergy optimization methods. It deals with problems in which a flowing fluid transfers heat and mass from a point to a surface or volume. The constructal law describes the possible optimum shapes and configurations in simple connecting tree-shaped distribution systems (networks) with the least amount of irreversibility. Zamfirescu and Bejan [19] proposed a tree-shaped structure as a CTES based on constructal law. Ice production on parallel plates and parallel cylinders has been studied, and an optimized shape structure has been suggested. Temperature difference, pressure drop, storage time, and material type have been considered as constrains. Ice production per unit volume is optimized for the mentioned assumptions as constrains. Bejan et al. [18] studied a methodology of how to optimize a thermal energy storage (TES) system according to a thorough first and second law analyses and economical considerations. In other words, a proper geometry selected for a CTES affects the overall performance of the system and can be optimized in different ways by considering the design

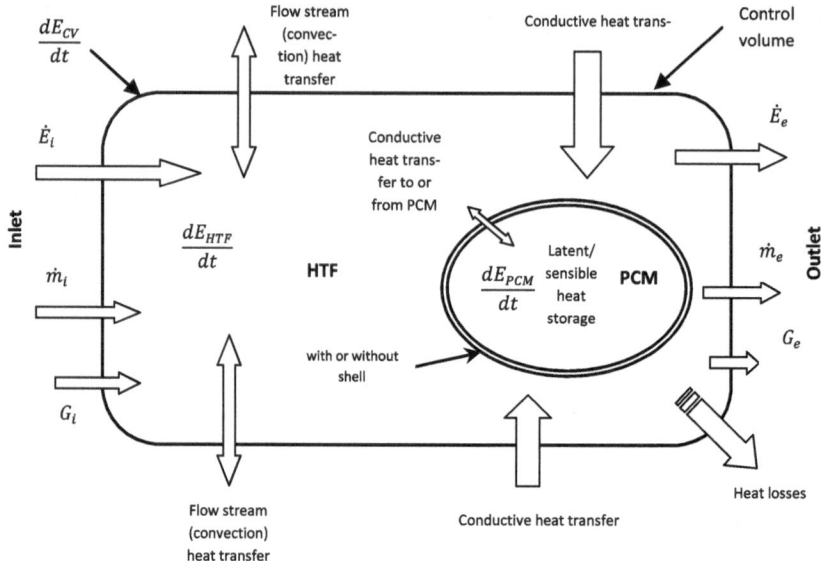

Fig. 1.2 Numerical model based on mass, heat, and entransy transfer balance for a typical control volume surrounded by flowing (or still) *HTF* and *PCM* inside a storage tank

parameters as a group of constraint factors. A good knowledge of heat transfer, fluid flow, and thermodynamics for selecting a cost-effective way of cold storage is necessary for any new geometry design.

1.3 Exergy Versus Energy Analysis

Energy analysis gives an overall view of system efficiency. Heat transfer and heat loss rates are incorporated into these performance efficiency factors, but there are no consideration about the entropy generation and the importance of the surrounding and environmental conditions. That is why the second law analysis is much more important to assess the lost available work according to the value of the exergy destruction and exergy deliveries to the system.

Bouzid et al. [20] investigated a fully developed laminar ice slurry flow entropy generation analytically. A non-Newtonian power-law model has been utilized to simulate the slurry flow. The relation between the entropy generation and three main variables (ice mass fraction and two dimensionless groups) has been found. Any ice mass fraction increment causes the velocity profile to be more flattened and consequently more entropy generation in the slurry flow.

MacPhee and Dincer [21] considered a CTES consisting of special ice capsules filled in a horizontal cylindrical tank. Two different temperature distributions (linear and quadratic) have been considered to model the CTES system analytically. A fully developed plug flow has been also considered with a mean porosity in a steady-state energy equation. One- and two-dimensional conduction and their importance have been evaluated. The effect of radial conduction found more important than the effect of axial conduction in the porous medium tank. Temperature distribution in the HTF has been evaluated, and the first and more importantly the second law analysis have been conducted to assess the CTES performance. In this study, heat loss through the wall of a tank has been considered and a flow exergy analysis of these CTES tanks has been recommended for any industrial applications.

In case of thorough thermodynamic analysis, especially according to exergy efficiencies of each component of a CTES system integrated to a refrigeration or A/C equipment, the portion of irreversibility of each component is found and later improvement of that individual system can be considered through using more efficient techniques from design to manufacturing. Rosen and Dincer [22] investigated some of important issues to be considered in performance of a sensible cold storage tank through exergy analysis which lead to more meaningful recommendations to improve the energy management and to reduce the systems energy consumptions. By reducing the heat losses of the tank, using the advantage of stratification inside the storage tank and avoiding mixing of the layers and finally by using more efficient pumps, the overall performance of the CTES system would be increased. Bakan et al. [23] investigated some factors on an ethylene glycol sensible thermal storage system capable of storing cold in a 3,50,000 kg tank. Chiller COP highly depends on storage and ambient temperature. The performance is affected by the chiller COP, HTF mass flow rate, and heat losses. A mild dependence on ambient temperature to the heat loss has been also verified.

Yamaha et al. [24] defined a response-based energy efficiency of the charging and discharging intervals accounting the outlet temperature of two different CTES systems (ice-on-coil and dynamic type ice generator). For the ice-on-coil storage system, HTF variable has been found as the most important parameter affecting the performance of the unit. Different inlet temperatures of the HTF and an added mixing process in the storage tank have also direct effects on the efficiency of the system.

In most thermal analysis of a TES system, there are two different approaches to assess the performance of the system. In most cases in analytical and numerical methods, a uniform distribution of PCM in HTF has been considered and the equations of transport phenomena have been developed considering a porous medium approach in one or two dimensions. The energy interactions between the PCM and the HTF are modeled by use of some experimental average formulations. In the other approach, a CFD code or package is used to find out the temperature and the velocity vector in two or three dimensions in a heat transfer fluid domain. As it is usually difficult to model a TES tank due to complicated geometry, symmetric patterns will help to reduce the magnitude of numerical experiment. So, the chosen cell usually has a repeatable pattern. All the boundary and initial conditions should be selected in a way as close as possible to the TES tank and PCM conditions.

The latter approach is very promising as powerful software packages are available today with substantial computer hardware memory and storages. So, it is better to obtain the velocity vector, temperature, and pressure variable and construct the corresponding fields for thermal assessment procedure.

In CTES systems for HVAC applications, spherical capsules are usually used in different diameters to fill a TES tank (usually in cylindrical shapes). These capsules are filled with distilled water. There are two different capsule arrangements available for spherical capsules in the numerical modeling. It can be a random arrangement of the capsules which is usually used in porous medium approach with a uniform porosity (as a function of capsule diameter to tank diameter) or a repeatable arrangement of the capsules.

There are some limitations in practical applications such as pumping power, HTF and PCM phase change temperature difference, specific thermal capacity of HTF and latent capacity of the PCM, ambient temperature where it is needed to store energy. For each of these, there is a range of values necessarily not a very wide range according to the type of use and applicability. For instance, in HVAC systems, distilled water is usually used as PCM encapsulated in some type of plastic/PVC containers and water/ethanol mixture is used as HTF. So, it would be beneficial to improve the fluid flow domain and enhance some of the overall thermal properties of the system before any economic and thermodynamic assessment. Entransy dissipation and synergy (coordination) between the velocity and the gradient temperature fields can be used as two thermal assessment tools to enhance the thermal performance of a thermal system. They will help the researchers and designers to optimize the system for a particular task or choose the best if there are some different available design options.

1.4 Entransy Extremum Theory

Qun and Jianxun proposed a weighted temperature, temperature difference, and equivalent thermal resistance for a convection optimization method [25]. An equivalent minimum thermal resistance theory which is another representation of entransy dissipation extremum theory has been developed, and a number of test cases have been solved to show the applicability of the method. The equivalent weighted thermal resistance is found from the ratio of equivalent temperature difference and the total heat convected through an open system. In this method, convective heat transfer is optimized by thermal resistance minimization when a fixed temperature difference or heat flux is imposed to the thermal system. A normalized temperature difference and thermal resistance have been also proposed to consider for cases where the temperature of the boundaries are imposed by the heat transfer assumptions. In these cases, a normalized resistance would be a better representative variable for optimization.

Xu showed that the heat conduction Fourier equation and Navier–Stokes fluid flow equations can be derived from a variational principle applied on entransy balance equation of an infinitesimal element as an open system [26]. Entransy of the net heat transfer to the system and also entransy of the net work done on the system have been used to formulate an equation and using variational technique to find the heat conduction and fluid flow equations. A volume integral of the net flow of entransy on the element has been used to be optimized with Euler–Lagrange equation. It would be one of the capabilities of entransy analysis as it is not expected from an entropy minimization point of view.

Cheng and Liang proposed a "maximum principle of entransy loss" concept as an equivalent method of conventional thermodynamic optimizations [27]. They showed that for reversible thermodynamic cycles, the maximum possible work output is equal to the entransy dissipation as there is no temperature difference between the heat sources with the working fluid (reversible cycle). The entropy generation is zero for these types of cycles, but it is possible to predict the maximum work output based on the high temperature heat transfer medium and its temperature. The work output is increased in case of higher values of high temperature source heat transfer rate at higher temperatures and lower values of cold reservoir temperature. It is also proposed as a better tool to assess the performance of irreversible cycles such as Brayton cycle which has been discussed in details.

Any TES system can be regarded as a heat exchanger and some of the terminology used would be the same such as NTU (number of transfer unit) and thermal resistance.

References

1. Guo Z, Zhu H, Liang X (2007) Entransy—a physical quantity describing heat transfer ability. Int J Heat Mass Transfer 50:2545–2556
2. Meng J, Liang X, Li Z (2005) Field synergy optimization and enhanced heat transfer by multi-longitudinal vortexes flow in tube. Int J Heat Mass Transfer 48:3331–3337
3. Guo Z, Liu X, Tao W, Shah R (2010) Effectiveness–thermal resistance method for heat exchanger design and analysis. Int J Heat Mass Transfer 53:2877–2884
4. Chen L, Chen Q, Li Z, Guo Z (2009) Optimization for a heat exchanger couple based on the minimum thermal resistance principle. Int J Heat Mass Transfer 52:4778–4784
5. Chen Q, Wang M, Pan N, Guo Z (2009) Optimization principles for convective heat transfer. Energy 34:1199–1206
6. Jia H, Liu W, Liu Z (2012) Enhancing convective heat transfer based on minimum power consumption principle. Chem Eng Sci 69:225–230
7. Xiao Q, Chen L, Sun F (2011) Constructal entransy dissipation rate minimization for a heat generating volume cooled by forced convection. Chin Sci Bull 56:2966–2973
8. Liu W, Liu Z, Jia H, Fan A, Nakayama A (2011) Entransy expression of the second law of thermodynamics and its application to optimization in heat transfer process. Int J Heat Mass Transfer 54:3049–3059
9. Chen L (2010) Constructal entransy dissipation minimization for 'volume-point' heat conduction. J Phys D Appl Phys 41:1075–1088

10. Wu J, Liang X (2008) Application of entransy dissipation extremum principle in radiative heat transfer optimization. Sci China Ser E Technol Sci 51:1306–1314
11. Tao W, Guo Z, Wang B (2002) Field synergy principle for enhancing convective heat transfer—its extension and numerical verifications. Int J Heat Mass Transfer 45:3849–3856
12. Guo Z, Li D, Wang B (1998) A novel concept for convective heat transfer enhancement. Int J Heat Mass Transfer 41:2221–2225
13. Guo Z, Tao W, Shah R (2005) The field synergy (coordination) principle and its applications in enhancing single phase convective heat transfer. Int J Heat Mass Transfer 48:1797–1807
14. Guo J, Xu M, Cheng L (2009) The application of field synergy number in shell-and-tube heat exchanger optimization design. Appl Energy 86:2079–2087
15. Regin AF, Solanki SC, Saini JS (2009) An analysis of a packed bed latent heat thermal energy storage system using PCM capsules: numerical investigation. Renewable Energy 34:1765–1773
16. Amri N, Amin M, Belusko M, Bruno F (2009) Optimisation of a phase change thermal storage system. Eng and Tech 3:708–712
17. Akif M, Ozdogan M, Gunerhan H, Hepbasli A (2010) Energetic and exergetic analysis and assessment of a thermal energy storage (TES) unit for building applications. Water 42:1896–1901
18. Bejan A, Tsatsaronis G, Moran M (1995) Thermal design and optimization. Wiley, New york
19. Zamfirescu C, Bejan A (2005) Tree-shaped structures for cold storage. Int J Refrigeration 28:231–241
20. Bouzid N, Saouli S, Saouli SA (2008) Entropy generation in ice slurry pipe flow. Int J Refrigeration 31:1453–1457
21. Macphee D, Dincer I (2009) Thermal modeling of a packed bed thermal energy storage system during charging. Appl Thermal Eng 29:695–705
22. Rosen M, Dincer I (2003) Exergy methods for assessing and comparing thermal storage systems. Int J Energy Res 27:415–430
23. Bakan K, Dincer I, Rosen MA (2008) Exergoeconomic analysis of glycol cold thermal energy storage systems for building applications. Int J Energy Res 32:215–225
24. Yamaha M, Nakahara N, Chiba R (2008) Studies on thermal characteristics of ice thermal storage tank and a methodology for estimation of tank efficiency. Int J Energy Res 32:226–241
25. Chen Q, Ren J (2008) Generalized thermal resistance for convective heat transfer and its relation to entransy dissipation. Chin Sci Bull 53:3753–3761
26. Xu M (2012) Variational principles in terms of entransy for heat transfer. Energy 44:973–977
27. Cheng X, Liang X (2012) Entransy loss in thermodynamic processes and its application. Energy 44:964–972

Chapter 2
Fundamentals of Entransy and Entransy Dissipation Theory

Abstract The entransy is a new developed parameter that is effective in opti-mization of heat transfer. It can be used as an evaluation of the transport ability of heat. In this chapter, based on the energy conservation equation, the entransy balance equations for heat conduction and convective heat transfer are developed. The entransy dissipation extreme principle is developed. This extreme principle can be concluded into the minimum thermal resistance principle defined by en-transy dissipation.

Keywords Energy conservation · Entransy dissipation · Extreme principle · Heat transfer · Thermal resistance · Conduction · Convection

In recent years, with the increasing of the living standard, the global increasing consumption of limited primary energy has become such a big concern not just for the lack of the primary energy, but for the depletion of the ozone layer and global warming. And also, it has been estimated that more than 80 % of the worldwide energy utilization involves the heat transfer process. Thermal engineering has for a long time recognized the huge potential for conserving energy and decreasing CO_2 emission so as to reduce the global warming effect through efficient heat transfer techniques. In general, approaches for heat transfer enhancement have been explored and employed over the full scope of energy generation, conversion, consumption, and conservation. Design considerations to optimize heat transfer have often been taken as the key for better energy utilization and have been evolving into a well-developed knowledge branch in both physics and engineering.

During the last several decades, a large number of heat transfer enhancement technologies have been developed, and they have successfully cut down not only the energy consumption, but also the cost of equipment itself. However, com-paring with other scientific issues, engineering heat transfer is still considered to be an experimental problem and most approaches developed are empirical or semi-empirical with no adequate theoretical base. For instance, for a given set of constraints, it is nearly impossible to design a heat-exchanger rig with the optimal heat transfer performance so as to minimize the energy consumption.

J. Gu and Z. Gan, *Entransy in Phase-Change Systems*,
SpringerBriefs in Thermal Engineering and Applied Science,
DOI: 10.1007/978-3-319-07428-3_2, © The Author(s) 2014

Therefore, scientists developed several different theories and methods to optimize heat transfer, such as the constructal theory and the second law of thermodynamics theory in terms of entropy. The second law of thermodynamics is one of the most important fundamental laws in physics, which originates from the study of the efficiency of heat engine and places constraints upon the direction of heat transfer and the attainable efficiencies of heat engines. The concept of entropy introduced by Clausius for mathematically describing the second law of thermodynamics has stretched this law across almost every discipline of science. However, in the framework of the classical thermodynamics, the definition of entropy is abstract and ambiguous, which was noted even by Clausius. This has induced some controversies for statements related to the entropy. Recently, Bertola and Cafaro found that the principle of minimum entropy production is not compatible with continuum mechanics. Herwig showed that the assessment criterion for heat transfer enhancement based on the heat transfer theory contradicts the ones based on the second law of thermodynamics. The entropy generation number defined by Bejan is not consistent with the exchanger effectiveness, which describes the heat exchanger performance [1]. Shah and Skiepko found that the heat exchanger effectiveness can be maximum, minimum, or in between when the entropy generation achieves its minimum value for 18 kinds of heat exchangers, which does not totally conform to the fact that the reduction of entropy generation leads to the improvement of the heat exchanger performance. These findings signal that the concepts of entropy and entropy generation may not be perfect for describing the second law of thermodynamics for heat transfer.

Although there has been effort to modify the expression of the second law of thermodynamics and to improve the classical thermodynamics by considering the Carnot construction cycling in a finite time, the eminent position of entropy in thermodynamics has not been questioned. Recently, Guo et al. introduced the concepts of entransy and entransy dissipation to measure, respectively, the heat transfer capacity of an object or a system, and the loss of such capacity during a heat transfer process. Moreover, Guo et al. [2] defined two new physical quantities called entransy and entransy dissipation for describing the heat transfer ability and irreversibility of heat conduction, respectively. Guo et al. have introduced a dimensionless method for the entransy dissipation and defined an entransy dissipation number, which can serve as the heat exchanger performance evaluation criterion. Based on the concept of entransy dissipation, an equivalent thermal resistance of heat exchanger was defined, which is consistent with the exchanger effectiveness [3] and consequently, developed the minimum entransy dissipation-based thermal resistance principle to optimize the processes of heat conduction [2, 4], convective heat transfer [5–7], thermal radiation, and in heat exchangers [3].

2.1 The Definition of Entransy and Entransy Dissipation

Guo et al. found that all transport processes contain two different types of physical quantities due to the existing irreversibility: the conserved ones and the nonconserved ones, and the loss or dissipation in the nonconserved quantities can then be used as the measurements of the irreversibility in the transport process. Taking an electric system as an example, although both the electric charge and the total energy are conserved during an electric conduction, the electric energy, however, is not conserved and it is partly dissipated into the thermal energy form due to the existence of the electrical resistance. Consequently, the electrical energy dissipation rate is often regarded as the irreversibility measurement in the electric conduction process. Similarly, for a viscous fluid flow, both the mass and the momentum of the fluid, transported during the fluid flow, are conserved, whereas the mechanical energy, including both the potential and kinetic energies, of the fluid is turned partially into the thermal energy form due to the viscous dissipation. As a result, the mechanical energy dissipation is a common measure of irreversibility in a fluid flow process. The above two examples show that the mass, or the electric quantity, is conserved during the transport processes, while some form of the energy associated with them is not. This loss or dissipation of the energy can be used as the measurement of irreversibility in these transport processes. However, an irreversible heat transfer process seems to have its own particularity, for the heat energy always remains constant during transfer and it does not appear to be readily clear what the nonconserved quantity is in a heat transfer process. Based on the analogy between electrical and heat conductions, Guo et al. made a comparison between electrical conduction and heat conduction as shown in Table 2.1 [2].

It could be found in the table that there is no corresponding parameter in heat conduction for the electrical potential energy in a capacitor, and hence they defined an equivalent quantity, G, that corresponding to the electrical potential energy in a capacitor, which is called "**entransy.**" They further derived Eq. (2.1) according to the similar procedure of the derivation of the electrical potential energy in a capacitor. Entransy was originally referred to as the heat transport potential capacity in an earlier paper by the Guo et al. [2].

$$G = \frac{1}{2}QT[\text{JK}] \qquad (2.1)$$

where Q is, $mc_v T$, the thermal energy or stored heat in a body at constant volume kept at temperature T. It is equivalent to potential electrical energy in a capacitor, which makes a current (heat flow) between two objects connected with a resistance together at two different potential levels. It is also possible to explain a heat transfer process through the analogy as depicted in Fig. 2.1.

Entransy represents the heat transfer ability of an object. It possesses both the nature of "energy" and the transfer ability. If an object is put in contact with an infinite number of heat sinks that have infinitesimally lower temperatures, the total quantity of "potential energy" of heat, whose output can be $\frac{1}{2}QT$. Biot suggested a

Table 2.1 Electrical and thermal energy variable and equation similarities

Electrical energy	Current	Resistance	Capacitance	Electrical potential (Voltage)	Current density	Ohm's law
Stored charge in a capacitor						
Q	I	R	C	U^a	$\dot{q}$	$\dot{q} = -K\frac{d\psi}{dn}$
C	$\frac{C}{s}$ (or A)	Ω	F	V	$\frac{C}{m^2 s}$	
Thermal energy	Heat flow	resistance	Heat capacity	Thermal potential (Temperature)	Heat flux	Fourier law
Stored heat in a body						
$Q = mc_v T$	$\dot{Q}$	R	$C = \frac{Q}{T}$	T^a	$\dot{q}$	$\dot{q} = -K\frac{dT}{dn}$
J	$\frac{J}{s}$ (or W)	$\frac{K}{W}$	$\frac{1}{K}$	K	$\frac{J}{m^2 s}$	

[a] Voltage and temperature are potential parameters and need a reference for measurement ($T = \Delta T$ and the difference measured from 0 K)

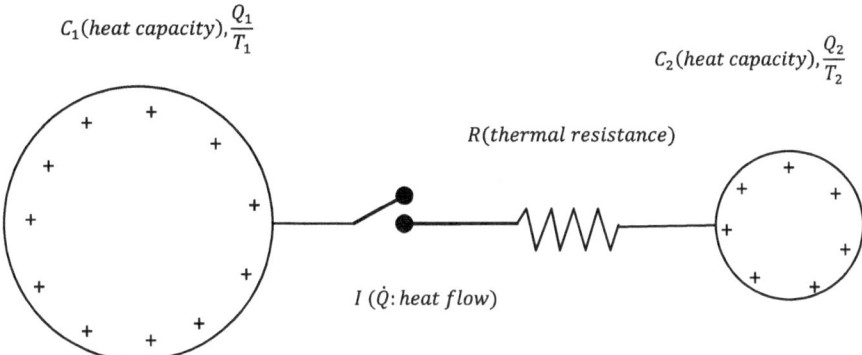

Fig. 2.1 Descriptive analogy between heat flow and electrical current with corresponding parameters and terminology in heat transfer (e.g., capacitor and resistance)

similar concept in the derivation of the differential conduction equation using the variation method. Eckert and Drake [8] pointed out that, Biot in a series of papers beginning in 1955, formulated from the ideas of irreversible thermal dynamics, a variational equivalent of the heat conduction equation that constituted a thermodynamical analogy to Hamilton's principle in mechanics and led to a Lagrangian formulation of the heat conduction problem in terms of generalized coordinates…, Biot defines a thermal potential as $E = {}^1\!/_2 \iiint_\Omega \rho c T^2 \mathrm{d}V$ … The thermal potential "E" plays a role analogous to a potential energy…

However, Biot did not further explain the physical meaning of thermal potential and its application was not found later except in the approximate solutions of anisotropic conduction problems. Accompanying the electric charge, the electric energy is transported during electric conduction. Similarly, along with the heat, the entransy is transported during heat transfer too. Furthermore, when a quantity of heat is transferred from a high temperature to a low temperature, the entransy is reduced and some of it is dissipated during the heat transport. The lost entransy is called entransy dissipation. Entransy dissipation is an evaluation of the irreversibility of heat transport ability.

For a transient heat conduction process without any heat transfer with ambient environment, the thermal energy conservation equation can be expressed as

$$\rho c_p \frac{\partial T}{\partial t} = \nabla \cdot (\lambda \nabla T) \tag{2.2}$$

where ρ, c_p, λ are density, constant pressure, specific heat and thermal conductivity, respectively. The entransy equilibrium equation can be obtained by multiplying both sides by temperature T:

$$\rho c_p T \frac{\partial T}{\partial t} = \nabla \cdot (\lambda T \nabla T) - \lambda |\nabla T|^2 \tag{2.3}$$

where the left side represents the time variation of the entransy stored per unit volume, the first term on the right side is the entransy transfer from one object to another associated with heat transfer, while the second term is the local entransy dissipation rate due to heat conduction. This is similarly like electric energy dissipation during an electric conduction process or mechanical energy dissipation during mechanical moving process. Since electrical energy and mechanical energy dissipations are both irreversibility measure of their respective process, entransy dissipation rate is hence a measure of the irreversibility in heat transfer process and can be written as

$$\dot{G} = \lambda |\nabla T|^2. \tag{2.4}$$

In a similar manner, the thermal energy conservation equation for a steady-state convective heat transfer process with no heat source can be expressed,

$$\rho c_p U_f \cdot \nabla T = \nabla \cdot (\lambda \nabla T), \tag{2.5}$$

where U_f is the velocity vector of the fluid. Similarly, the entransy equilibrium equation for the convective heat transfer can be derived by multiplying both sides by temperature T,

$$U_f \cdot \nabla \left(\frac{\rho c_p T^2}{2} \right) = \nabla \cdot (\lambda T \nabla T) - \lambda |\nabla T|^2 \tag{2.6}$$

The left side of this equation express the entransy transferred associated with the fluid particles motion, while the right side of this equation is in the same form as the heat conduction process, which include the entransy diffusion within the fluid due to temperature gradient and the local entransy dissipation rate.

By integrating Eq. (2.6) over the entire domain, transforming the volume integral to the surface integral on the domain boundary, and ignoring the heat diffusion in the flow direction at both inlets and outlets, we can obtain

$$\left(\frac{1}{2} \rho \dot{V} c_p T^2 \right)_{\text{out}} - \left(\frac{1}{2} \rho \dot{V} c_p T^2 \right)_{\text{in}} = \iint_{\Gamma} \vec{n} \cdot \lambda T \nabla T \mathrm{d}A - \iiint_{\Omega} \lambda |\nabla T|^2 \mathrm{d}V \tag{2.7}$$

where the first term on the left side describes the entransy flowing out of the domain, while the second term is the entransy flowing into the domain. On the right side, the first term indicates the entransy flow rate induced by heat transfer through the domain boundary, while the second term can be viewed as the total entransy dissipation rate

$$\dot{G} = \iiint_{\Omega} \lambda |\nabla T|^2 \mathrm{d}V. \tag{2.8}$$

Similar to electrical resistance, the entransy dissipation-based thermal resistance of a heat exchanger is defined as

$$R_{ex} = \frac{\dot{G}}{Q^2},$$ (2.9)

where R_{ex} is the entransy dissipation-based thermal resistance, $\dot{G}$ is the entransy dissipation rate during the heat transfer process and Q is the total heat transfer rate.

2.2 Entransy Analysis in Conduction Heat Transfer

The conduction heat transfer equation with no heat source available in a domain is

$$\rho c_v \frac{dT}{dt} = \nabla \cdot (K \nabla T)$$ (2.10)

By multiplying both sides by T, the entransy equation is

$$\rho c_v T \frac{dT}{dt} = -\nabla \cdot (TK\nabla T) + K\nabla T \cdot \nabla T,$$ (2.11)

or

$$\frac{dG}{dt} = -\nabla \cdot \dot{G} - G_\varphi,$$ (2.12)

where G is entransy density, $\frac{1}{2}\rho c_v T^2$, G_φ is entransy dissipation as $K(\nabla T)^2$, and $\dot{G}$ is entransy flux. The presence of temperature gradient to the power of two in local entransy dissipation resembles viscose dissipation term in entropy generation in fluid flow, which is similar to electrical dissipation. For all these squared terms, energy should be paid (consumed) to maintain the corresponding flow (e.g., heat or mass). For instance, to maintain a viscous flow, pumping power should be provided to have the fluid flowing. For heat transfer, energy should be delivered to maintain the heat flow as there is always a resistance present and entransy dissipation. It looks like a connecting concept between heat transfer and thermodynamics and furthermore between different energy and heat transfer modes.

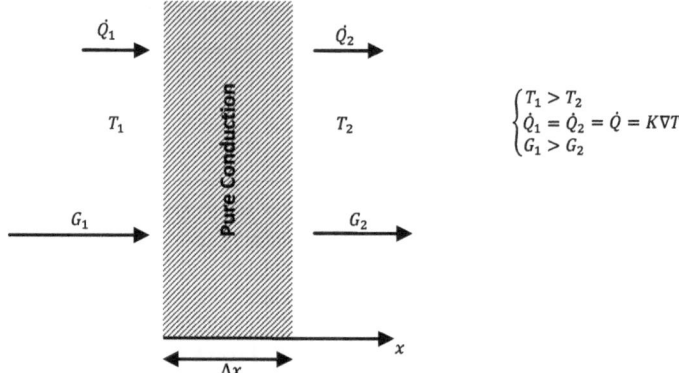

Fig. 2.2 Steady heat conduction in an object with very large depth and height (one-dimensional conduction)

2.3 Equivalent of Thermal Resistance in Heat Convection and Entransy Dissipation

In a one-dimensional or infinitesimal differential heat conduction case, thermal resistance at a point is easy to define as $R = \Delta T / \dot{Q}$ On the other hand, for multi-dimensional heat conduction case, it is not straightforward to calculate it and needs some definitions and averaging procedure. Now, it is possible to define an equivalent thermal resistance based on entransy dissipation, which occurs in thermal resistance

$$R = \frac{G_\varphi}{\dot{Q}^2} \qquad (2.13)$$

where $G_\varphi = \int_V K(\nabla T)^2 dV$ is the rate of entransy dissipation over a volume has a temperature gradient (thermal potential gradient) with a heat flow rate of $\dot{Q}$ caused by ∇T established in the volume in heat conduction mode. The equivalent thermal resistance can be found in any arbitrary shape in a three-dimensional and general case with no limitations. In a one-dimensional conduction case, it will be trivial as

$$R = \frac{\dot{Q} \, \Delta T}{\dot{Q}^2} = \frac{\Delta T}{\dot{Q}} \qquad (2.14)$$

The temperature gradient in a one-dimensional Fourier heat conduction case across a very long and wide solid bar with a thickness of Δx can be found where a constant heat flux of $\dot{Q}$ transfer thermal heat energy from higher temperature to the colder side across the object shown in Fig. 2.2. Energy is conserved and $\dot{Q}$ is constant for this special one-dimensional case, but there is entransy dissipation as the thermal resistance is available to facilitate the heat flux (equivalent to current in electrical energy). Entransy leaving the colder side, G_2 is less than G_1 as a result of entransy dissipation G_φ which can be found as follows:

$$G_1 - G_2 = G_\varphi = -\int_0^{\Delta x} K\nabla T.\nabla T \mathrm{d}x = -\int_0^{\Delta x} \dot{Q}\frac{\mathrm{d}T}{\mathrm{d}x}\mathrm{d}x = \dot{Q}(T_1 - T_2) \qquad (2.15)$$

The heat flow direction is always in an entransy dissipation mode, which means G_φ should be positive (another interpretation of the second law of thermodynamics).

2.4   Conclusions

Entransy is a parameter developed in recent years. It is effective in optimization of heat transfer. Entransy is an evaluation of the transport ability of heat. Both the amount of heat and the potential contribute to the entransy. Entransy will be lost during heat transportation from a high to a low temperature, and entransy dissipation will be produced.

In this chapter, based on the energy conservation equation, the entransy balance equations for heat conduction and convective heat transfer have been developed. The entransy dissipation extreme principles are also developed, that is, the maximum entransy dissipation corresponds to the maximum heat flux for prescribed temperature difference and the minimum entransy dissipation corresponds to minimum temperature difference for prescribed heat flux. This extreme principle can be concluded into the minimum thermal resistance principle defined by entransy dissipation.

References

1. Guo J, Xu M, Cheng L (2009) The application of field synergy number in shell-and-tube heat exchanger optimization design. Appl Energy 86: 2079–2087
2. Guo Z, Zhu H, Liang X (2007) Entransy—A physical quantity describing heat transfer ability. Int J Heat Mass Transfer 50:2545–2556
3. Meng J, Liang X, Li Z (2005) Field synergy optimization and enhanced heat transfer by multi-longitudinal vortexes flow in tube. Int J Heat Mass Transfer 48:3331–3337
4. Chen L, Chen Q, Li Z, Guo Z (2009) Optimization for a heat exchanger couple based on the minimum thermal resistance principle. Int J Heat Mass Transfer 52:4778–4784
5. Meng J, Liang X, Li Z (2005) Field synergy optimization and enhanced heat transfer by multi-longitudinal vortexes flow in tube. Int J Heat Mass Transfer 48:3331–3337
6. Chen Q, Wang M, Pan N, Guo Z (2009) Optimization principles for convective heat transfer. Energy 34:1199–1206
7. Chen Q, Ren J (2008) Generalized thermal resistance for convective heat transfer and its relation to entransy dissipation. Chin Sci Bull 53:3753–3761
8. Eckert ERG, Drake RM (1959) Heat and mass transfer. McGraw-Hill Inc, New York

Chapter 3
Application of Entransy Theory in Thermal Storage System

Abstract There are many possible capsule and tank geometry types and arrangements. It is very helpful to assess the performance of capsule arrangements, shapes, and flow field configurations. For this purpose, a general tool will be needed to assess the heat transfer which takes place between the capsules and the HTF and evaluate how efficient the process is. Then any shape and configuration can be assessed and thus help the designers and manufacturers offer more energy saving TES systems. In this chapter, thermal assessment parameters for TES system performance are introduced including nondimensional TES equivalent thermal resistance, equivalent temperature (potential) difference, number of transfer units (NTU), and synergy density through a heat exchanger approach.

Keywords Efficiency · Exergy · Synergy · Phase change · Thermal energy storage · Number of transfer units · Heat exchanger

In order to study and formulate thermal energy storage and its equivalent thermal resistance through an entransy analysis, a simple system has been considered as shown in Fig. 3.1. A simple heat exchanger (e.g., a straight pipe) has been immersed in a constant temperature bath (at temperature θ). The bath is well insulated and the constant temperature is provided through a proper phase change process at temperature θ. The heat exchanger wall is considered at this constant temperature. A heat transfer fluid (HTF) flows inside the pipe and enters the system at temperature T_{in} which is considered higher than the bath temperature in this example. Energy conservation equation for the fluid flow inside the pipe is,

$$\dot{m}c\frac{\mathrm{d}T}{\mathrm{d}x} = -\dot{Q}_x(x), \tag{3.1}$$

where $\dot{m}$ is the mass flow rate, c is the heat capacity of the heat transfer fluid, and T is the fluid bulk temperature. Multiplying both sides of the Eq. (2.10) by T yields the following entransy balance equation after integration from inlet to exit of the pipe immersed in the constant temperature bath,

J. Gu and Z. Gan, *Entransy in Phase-Change Systems*,
SpringerBriefs in Thermal Engineering and Applied Science,
DOI: 10.1007/978-3-319-07428-3_3, © The Author(s) 2014

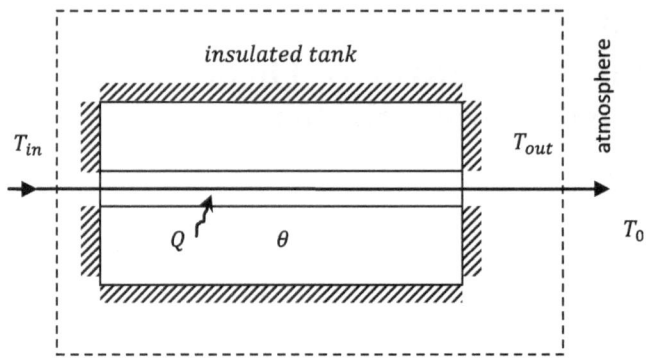

Fig. 3.1 A simplified insulated heat exchanger (e.g., a straight pipe) immersed in a constant temperature tank

$$G_{\text{out}} - G_{\text{in}} = \int\limits_{x_{\text{in}}}^{x_{\text{out}}} \dot{m}cT(x)\frac{\mathrm{d}T(x)}{\mathrm{d}x} = \frac{1}{2}\dot{m}cT_{\text{out}}^2 - \frac{1}{2}\dot{m}cT_{\text{in}}^2 \qquad (3.2)$$

As the temperature of the bath is lower than the fluid bulk temperature, heat transfer takes place from the fluid to the bath at constant bath temperature θ. Entransy dissipation in this convective heat transfer is

$$G_{\varphi} = G_{\text{in}} - G_{\text{out}} - \int\limits_{x_{\text{in}}}^{x_{\text{out}}} \theta\dot{Q}(x)\mathrm{d}x = \int\limits_{x_{\text{in}}}^{x_{\text{out}}} (T(x) - \theta)\dot{Q}(x)\mathrm{d}x \qquad (3.3)$$

In this convective heat transfer mode, it is possible to define an equivalent resistivity for the thermal energy storage

$$R_{\text{TES}} = \frac{G_{\varphi}}{\dot{Q}^2} = \frac{\int\limits_{x_{\text{in}}}^{x_{\text{out}}} (T(x) - \theta)\dot{Q}(x)\mathrm{d}x}{\left(\int\limits_{x_{\text{in}}}^{x_{\text{out}}} \dot{Q}(x)\mathrm{d}x\right)^2}. \qquad (3.4)$$

An equivalent temperature difference exists facilitating the convective heat transfer between the bulk fluid and the bath

$$\Delta T = R\dot{Q} = \frac{G_{\varphi}}{\dot{Q}} \qquad (3.5)$$

Bejan discussed the same heat exchanger and its entropy generation followed by internal and external irreversibility in detail [1].

3.1 Different Types of Performance Evaluation Coefficients: Efficiency and Effectiveness

There are some important coefficients to assess the performance of the storage system as for many other thermal engineering units. As the storage system acts as a heat exchanger, some of the terminologies used for them apply as well to CTES systems [2, 3]. Energetic and exergetic effectiveness coefficients are defined as the following ratios:

$$\varphi(t)\big|_{\text{energy(exergy)}} = \frac{\text{total energy(exergy)stored}}{\text{maximum possible energy(exergy)to be stored}} \qquad (3.6)$$

It can be expressed for charging or discharging processes. If it used for discharging processes, released energy values are inserted instead of stored energy values. The maximum possible energy (exergy) can be stored as the temperature decreases or increases (depending on the process—charging or discharging) from the initial to the inlet temperature for the storage tanks for instance.

The ratio of the delivered energy (exergy) to the amount of energy (exergy) which delivers to a CTES can be defined as energy or exergy efficiency for charging and discharging intervals

$$\eta = \frac{\text{usefuloutput energy}}{\text{required input}}\bigg|_{\text{discharging}}, \quad \frac{\text{energy stored}}{\text{inputenergy}}\bigg|_{\text{charging}}, \qquad (3.7)$$

$$\psi = \frac{\text{exergyrecovered}}{\text{exergy supplied}}\bigg|_{\text{discharging}}, \quad \frac{\text{exergystored}}{\text{exergy delivered}}\bigg|_{\text{charging}}, \qquad (3.8)$$

where, η and ψ denote the energy and exergy efficiencies, respectively. Higher values of these efficiencies and effectiveness mean a better performed CTES system according to the first and the second laws of thermodynamics. As the second law efficiencies include the surrounding conditions and the irreversibility responsible for entropy generation and consequently exergy destruction, they give better interpretations compared to the only energy analysis; therefore, they are more useful tools for decision-making processes.

There are three more parameters that can be also calculated for test cases comparison. A nondimensional equivalent thermal resistance, $R^+ = R/(mc)^{-1}$, number of transfer units (TES system as a heat exchanger), and synergy density (as the volume average of volume integral of synergy over the fluid domain) can be considered for comparing different examples of thermal systems.

3.2 Exergy Analysis

In order to evaluate all the required values for an exergy analysis, a study of entropy and entropy generation due to internal (within the control volume) and external (due to interactions of the system with its surroundings) irreversibility of a control volume is needed. Entropy rate balance for a control volume is

$$
\left\{ \begin{array}{c} \text{Instantanious rate} \\ \text{of entropy change} \\ \text{in a control volume} \end{array} \right\} = \left\{ \begin{array}{c} \text{Instantanious rates of} \\ \text{entropy transfer through} \\ \text{heat and mass flow} \end{array} \right\} + \left\{ \begin{array}{c} \text{Instantanious rate} \\ \text{of entropy} \\ \text{generation} \end{array} \right\}, \tag{3.9a}
$$

and its integral form can be expressed as

$$
\frac{\mathrm{d}S_{\mathrm{CV}}}{\mathrm{d}t} = \sum_j \int_{A_j} \frac{\dot{q}_j}{T_j} \mathrm{d}A_j + \sum_i \left(\int s\rho U_n \mathrm{d}A \right)_i - \sum_e \left(\int s\rho U_n \mathrm{d}A \right)_e + \dot{\sigma}_{\mathrm{CV}}, \tag{3.9b}
$$

where S_{CV} is the entropy of the control volume at time t, $\dot{q}_j$ is the amount of heat transfer rate across A_j at temperature T_j, s is the specific entropy, and $\dot{\sigma}_{\mathrm{CV}}$ is the instantaneous rate of entropy production due to internal and external irreversibilities for the control volume.

The exergy rate balance of the control volume can be expressed as

$$
\frac{\mathrm{d}\Xi_{\mathrm{CV}}}{\mathrm{d}t} = \sum_j \int_{Aj} \left(1 - \frac{T_0}{T_j} \right) \dot{q}_j \mathrm{d}A_j - \left(\dot{W}_{\mathrm{CV}} - P_0 \frac{\mathrm{d}V}{\mathrm{d}t} \right)
$$
$$
+ \sum_i \left(\int e_f \rho U_n \mathrm{d}A \right)_i - \sum_e \left(\int e_f \rho U_n \mathrm{d}A \right)_e + T_0 \dot{\sigma}_{\mathrm{CV}}, \tag{3.10}
$$

where, T_0 and P_0 denote the dead state (environment) temperature and pressure, respectively. Equation (2.7) can be expressed in a more compact notation as

$$
\frac{\mathrm{d}\Xi_{\mathrm{CV}}}{\mathrm{d}t} = \dot{\Xi}_q - \dot{\Xi}_W + \sum \dot{\Xi}_{fi} - \sum \dot{\Xi}_{fe} - \dot{\Xi}_d, \tag{3.11}
$$

where $\dot{\Xi}_q$ and $\dot{\Xi}_W$ are exergy transfer rates according to the heat and work interactions with the control volume, respectively. The quantity $\dot{\Xi}_d$ is the destruction of exergy rate due to entropy generated through the internal and external irreversibilities. The quantity $\dot{\Xi}_f$ denotes exergy flow rate which expressed by the following specific exergy (e_f) expression

$$e_f = h - h_0 + T_0(s - s_0) + \frac{U^2}{2} + gZ, \qquad (3.12)$$

where the zero subscript denotes the atmospheric (surrounding) conditions, h and s are the specific enthalpy and entropy content of the control volume, respectively [4].

3.3 Velocity and Temperature Gradient Fields' (Coordination) Synergy

In most of heat transfer enhancement techniques, there is a need to have more general definitions to find an enhancement degree or measure the improvement. Guo et al. [5] introduced a methodology to measure and compare convective heat transfer strength in different applications. It has opened numerous enhancement techniques for industries. In order to study convective heat transfer, they revisited a boundary layer convective heat transfer equation in a laminar flow for Fig. 3.2 as [6],

$$\rho c \left(u \frac{dT}{dy} + v \frac{dT}{dx} \right) = \frac{\partial}{\partial y} \left(K \frac{\partial T}{\partial y} \right). \qquad (3.13)$$

This equation has similar counterpart in a conductive heat transfer equation in a wall (Fig. 3.2) with uniform heat source distribution

$$-\dot{q} = \frac{\partial}{\partial y} \left(K \frac{\partial T}{\partial y} \right). \qquad (3.14)$$

It is interesting that the convective part of Eq. (3.13), $\rho c U \cdot \overrightarrow{\nabla T}$, acts like uniform heat source distribution in a conductive heat transfer mode. By integrating Eq. (3.13) over the boundary layer thickness, the wall heat flux is

$$\int_0^{\delta_t} \rho c \left(u \frac{dT}{dy} + v \frac{dT}{dx} \right) dy = -K \frac{\partial T}{\partial y} \bigg|_{\text{wall}} = -Q_{\text{wall}}, \qquad (3.15)$$

where δ_t is the thickness of the boundary layer. In heat transfer enhancement techniques (e.g., heat exchanger applications), it is desirable to increase wall heat flux. In such a case, the left-hand side of the above equation should be maximized within the fluid domain. In Eq. (3.14), by increasing heat source strength within the wall, heat transfer will be increased as well as the temperature gradient (right-hand side of the equation). So, by maximizing the integration of the left-hand side of the Eq. (3.15), heat transfer will be enhanced and consequently the wall heat flux will be increased. Equation 3.15 can also be written as

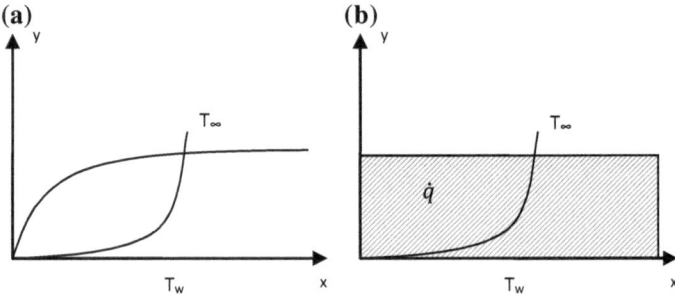

Fig. 3.2 Similarity between temperature profiles in **a** laminar boundary layer and **b** heat conduction with a uniform point heat source distribution

$$\int_0^{\delta_t} \rho c U \cdot \nabla T \mathrm{d}y = -\mathrm{K}\frac{\partial T}{\partial y}\bigg|_{\mathrm{wall}} = -Q_{\mathrm{wall}}. \qquad (3.16)$$

If the flow rate and the temperature difference of the wall and the bulk fluid are considered as almost relatively fixed values (technical limitations), it is possible to increase the heat transfer by increasing the value of $U \cdot \nabla T$. This scalar product of velocity vector and temperature gradient and its value is

$$U \cdot \nabla T = |U||\nabla T|\cos \beta, \qquad (3.17)$$

where β is the angle between these two vectors at each point in the domain. So the value of this product is a direct measure of the convective heat transfer or the convective strength. So by increasing $\cos \beta$ (configuration changes) it would be possible to have more heat transfer at fixed rates of mass flow and constant temperature difference between the inlet flow and the walls. It is also better to study the nondimensional interpretation of the equation as (on a flat plate for instance)

$$Nu_x = \mathrm{Re}_x\mathrm{Pr}\int_0^1 \overline{U} \cdot \nabla \overline{T}\mathrm{d}\bar{y}, \qquad (3.18)$$

where $\overline{U}$ is nondimensional velocity (U/U_∞) and U_∞ is the fluid bulk velocity, and $\overline{T}$ is the dimensionless temperature as $T/(T_\infty - T_w)$.

Field synergy principle proposed by Guo et al. [6] states: "the better synergy of velocity and temperature gradient/heat flow field, the higher the convective heat transfer rate under the same other conditions." In other words, fluid flow and heat flow fields' coordination would be a measurement new quantity which is called "synergy" for heat transfer strength. Tao et al. [7] expand the use of synergy in more general convective heat transfer equation which most applicable in engineering problems as

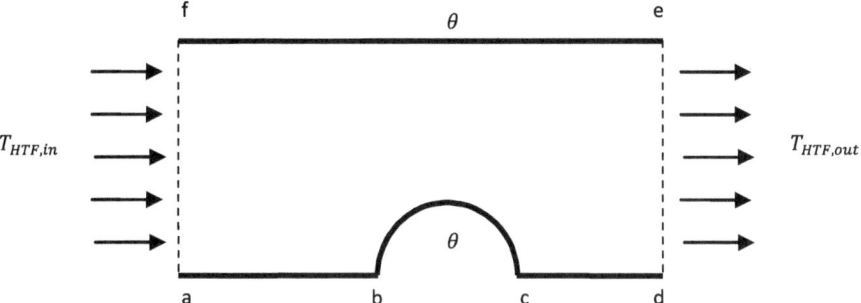

Fig. 3.3 Two dimensional convective heat transfer: a heat transfer fluid flow between two heating/or cooling walls at temperature θ

$$\rho c \left(u \frac{dT}{dy} + v \frac{dT}{dx} \right) = \frac{\partial}{\partial x} \left(K \frac{\partial T}{\partial x} \right) + \frac{\partial}{\partial y} \left(K \frac{\partial T}{\partial y} \right). \tag{3.19}$$

By integrating Eq. (3.19) on the two-dimensional domain defined in Fig. 3.3 and use of Gauss theorem, the following equation has been found:

$$\iint (\rho c U \cdot \nabla T) dx dy = \int_{abcd} \vec{n} \cdot K \overrightarrow{\nabla T} dl + \int_{de} \vec{n} \cdot K \overrightarrow{\nabla T} dl + \int_{ef} \vec{n} \cdot K \overrightarrow{\nabla T} dl + \int_{fa} \vec{n} \cdot K \overrightarrow{\nabla T} dl, \tag{3.20}$$

where $\vec{n}$ is the outward normal vector on the boundary and dl is the length differential on the boundaries. Equation 3.20 can be rearranged as

$$\iint (\rho c U \cdot \nabla T) dx dy - \int_{fa} \vec{n} \cdot K \overrightarrow{\nabla T} dl - \int_{de} \vec{n} \cdot K \overrightarrow{\nabla T} dl$$
$$= \int_{abcd} \vec{n} \cdot K \overrightarrow{\nabla T} dl + \int_{ef} \vec{n} \cdot K \overrightarrow{\nabla T} dl, \tag{3.21}$$

The right-hand side of the above equation calculates the heat transfer between the wall and the heat transfer fluid. In high Peclet number flow regimes (Pe > 100), the second and the third integrals on the left-hand side of the equation can be neglected as axial (normal to the inlet and outlet directions) fluid heat conduction relative to the fluid convection (first left-hand side term) and there would be the same concept for synergy interpretation which has been discussed earlier. Even in low Peclet number cases, higher synergy or better coordination between the flow and heat transfer felids enhances heat transfer; but synergy is not as important in high Peclet number flow fields.

Two more properties have been studied for thermal energy storage system performance assessment. Entransy and field synergy are two recently developed properties, especially over the past 6 years brought more meaning and introduce

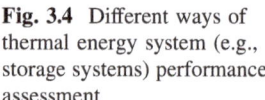
Fig. 3.4 Different ways of thermal energy system (e.g., storage systems) performance assessment

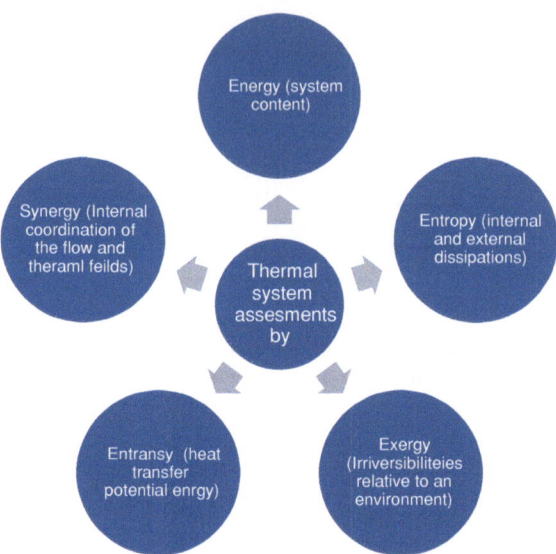

measurement criteria for thermal performance systems in a close relation to energy, entropy, and exergy. Figure 3.4 shows different thermal energy analyses based on first and second law of thermodynamics and two newly defined parameters (entransy and synergy).

3.4 Modeling

In order to solve a problem numerically, mass, momentum, and energy equations have been solved by ANSYS™ Benchmark (Fluent solver) software within the numerical domain. Three-dimensional geometry of each flow domain has been imported and mesh generation procedure has been done individually on the geometry. The flow is considered as steady state laminar fluid flow with no thermal viscous dissipations in thermal energy equation.

3.5 Applications to Thermal Energy Storage of Phase Change Materials

Heat exchanger design and performance assessment tools are applicable to thermal energy systems store sensible or latent heat. It is vital to do heat transfer and thermodynamic assessments on these heat exchangers to ensure how efficient they perform under different environmental and operational conditions.

In cold thermal energy storage applications for HVAC systems, there is usually an insulated tank filled with many capsules, typically spherical water capsules. A heat transfer fluid, a type of suitable alcohol (mixture) flows over the capsules to charge or discharge the stored thermal energy (sensible and latent heat). If the fluid flow temperature is well below the melting point of water, the water content of the capsules start solidifying and later on a higher temperature heat transfer fluid passing over the capsules to exchange heat with them and exit at a lower temperature for HVAC system use.

In most of the cases and under a typical fluid mass flow rate, there is a substantial period of time that all the capsules are under charging or discharging latent heat at the same time. So, it is a good assumption to consider a steady fluid flow over the capsules under a constant capsule wall temperature (melting temperature of the phase change material, water). The thermal resistance of the capsule wall imposes a small temperature change on two wall sides which is considered as negligible here.

A fluid cell of a TES system containing many spherical water/ice capsules has been selected. The cell is in contact to 4 (Group I) and 24 (Group II) capsules and a heat transfer fluid flow through the passage. For each test case, boundary conditions have been applied to the boundaries of the fluid domain mainly inlet, outlet, and wall of the capsules. There are four different types of boundary conditions used for the geometries. Inlet mass flow rates or uniform velocity distribution is always applied at the inlet. All the side walls except the spherical capsule walls are considered as symmetry boundaries as these walls of the domain are considered as symmetric planes between the cells in a large tank filled with the spherical capsules. The outlet boundary condition is applied at the exit. An extended length of symmetrical planes is considered for proper outlet boundary conditions at the outlet for Group II geometry. All the spherical surfaces are considered at the phase change temperature of the PCM (253.15 K for water in all test cases).

3.5.1 Assumptions

In order to solve the momentum and energy equations, the following assumptions have been considered for all the test cases in the next chapter:

- laminar and steady flow field,
- steady-state thermal analysis,
- constant HTF properties,
- constant wall temperature (phase change temperature),
- no buoyancy effect, and
- no thermal viscous dissipation.

Table 3.1 Geometry parameters of different test cases in Group I

	Number of complete capsules at the boundaries as walls	Inlet dimensions (m × m)	Inlet cross section area (m²)
Flow over 4 spherical capsules	1	0.08 × 0.08	0.0064

Table 3.2 Different diameter values used in Group I test cases

	d (m)
Flow over 4 spherical capsules	0.04, 0.08, 0.12

3.5.2 Heat Transfer Fluid Domain Dimensions

In order to perform numerical simulations and find the velocity and temperature fields, a typical geometry has been selected. A typical industrial TES tank consists of a large number of capsules (e.g., 2,500 spherical capsules per one cubic meter). It is difficult and time consuming to perform three-dimensional numerical modeling due to complicated geometry and randomness on a real TES tank contains the capsules. So, it is better to start with some elementary cell configurations which shape a repeatable element of the whole tank and using the symmetric geometries.

There are two different major streams in TES tank thermal modeling. The very first common way is to use porous medium approach (usually in one- and two-dimensional analyses). The other method is a numerical simulation on a repeatable geometry as the fluid domain in contact to capsule shapes in a tank. Careful attention needs in these kind of simulations to obtain accurate results such as using proper mesh generation techniques and analysis, using suitable convergence criteria, and finding a way to simulate inlet and outlet boundary conditions as close as possible to the actual (industrial) applications. These steps should be taken in any final thermal assessments and validations.

Two groups of numerical test cases have been considered. Tables 3.1 and 3.2 summarize the geometry parameter of each test in Group I. In this group, there are nine different tests run under different inlet conditions (Table 3.3).

Three-dimensional fluid flow domain of Group II is shown in Fig. 3.5. In this model, there are six complete spherical capsules on the boundaries as constant temperature walls (24 × (1/4) capsules). Tables 3.4 and 3.5 summarize geometry parameters for different test cases in Group II numerical simulations done under the inlet boundary conditions as in Table 3.6.

Table 3.3 Group I case numbers and its corresponding inlet temperature and mass flow rate

Inlet uniform velocity (m/s)	T_{in}(K)		
	268.15	263.15	258.15
Case I number			
5×10^{-4}	1	2	3
1×10^{-3}	4	5	6
2×10^{-3}	7	8	9

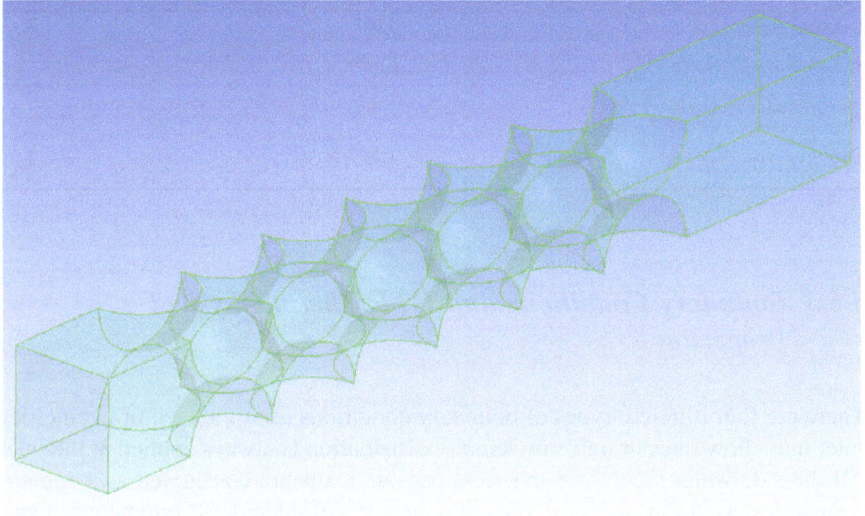

Fig. 3.5 Heat transfer fluid domain geometry in contact to 24 spherical capsules with an extended outlet length

Table 3.4 Geometry parameters of different test cases in Group II

	Number of complete capsules at the boundaries as walls	Inlet dimensions (m $\times$ m)	Inlet cross section area (m²)
Flow over 24 spherical capsules	6	0.08×0.08	0.0064

Table 3.5 Capsule diameter, enclosure length, constant wall area, and the HTF volume in each domain in Group II test cases

	d (m)	L (m)	Capsule surface area (m²)	HTF volume (m³)	Case number
Flow over 24 spherical capsules	0.08	0.72	0.48255	0.0030	Case II and III

Table 3.6 Inlet boundary conditions for two different test cases in Group II

Cases	$T_{in}(K)$	Inlet condition
Case IV	263.15	Inlet mass flow rate (kg/s) 0.003
Case V		Inlet uniform velocity (m/s) 0.0006

Table 3.7 Heat transfer fluid thermophysical properties

HTF—Ethanol (C_2H_6O)	Value	Unit
Density	789	$\frac{Kg}{m^3}$
Thermal conductivity	0.177	$\frac{W}{mK}$
Specific heat capacity	2,500	$\frac{J}{KgK}$
Dynamic viscosity	1.197×10^3	$\frac{Kg}{ms}$

3.5.3 Boundary Conditions and HTF Thermophysical Properties

There are four different types of boundary conditions used for each of geometries. Inlet mass flow rates or uniform velocity distribution is always applied at the inlet. All the side walls except the spherical capsule walls are considered as symmetry boundaries as these walls of the domain are considered as symmetric planes between the cells in a large tank filled with the spherical capsules. The outlet boundary condition is applied at the exit and an extended length at of symmetrical planes is considered for proper outlet boundary conditions at the outlet for some of the test cases (second series of test cases). All the spherical surfaces are considered at the phase change temperature of the PCM (253.15 K for water in all test cases). Thermophysical properties of Ethanol (C_2H_6O) are summarized in Table 3.7 as the HTF in all test cases.

3.5.4 Test Cases and the Results: Group I

Exergy efficiency for each test case has been calculated based on total heat transfer to the capsules at constant wall temperature (PCM phase change temperature). Figure 3.6 shows exergy efficiency of different test cases at different inlet conditions (inlet temperature and mass flow rate). Exergy efficiency is decreased by lowering the inlet temperature for all mass flow rates. High temperature difference between the inlet and phase change temperatures is corresponding to lower exergy

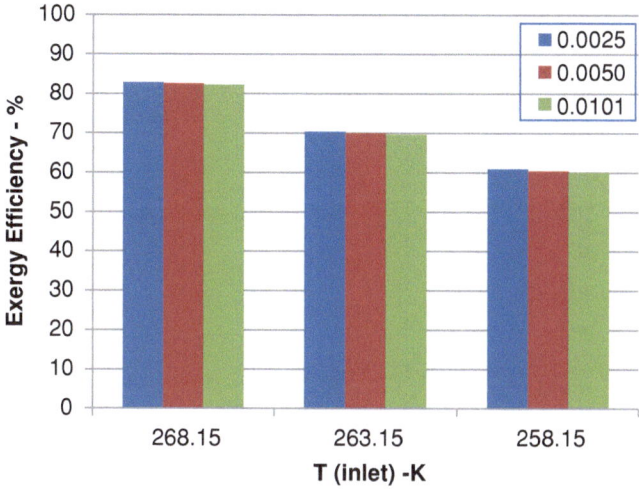

Fig. 3.6 Exergy efficiency at different inlet temperatures and flow rates

efficiency. Figure 3.7 shows exergy efficiencies at different inlet boundary conditions. Inlet temperature is considered the same. A higher ambient temperature for all the mass flow rates is corresponding to higher exergy efficiency.

In order to assess the overall performance of these arrangements as heat exchangers, entransy dissipation, thermal equivalent resistivity, heat transfer coefficient (1/R), and equivalent Nusselt number (based on equivalent heat transfer coefficient) are calculated for each of the cases at different inlet and found outlet temperatures and total heat transfer at the constant temperature walls. Figure 3.8 shows equivalent thermal resistivity. These values have been found independent of the inlet temperatures. Higher mass flow rates correspond to lower equivalent thermal resistivity. Figure 3.9 shows equivalent temperature difference versus wall-inlet temperature difference under Case I boundary conditions. A higher mass flow rate increases the equivalent temperature as a result of lower equivalent thermal resistance.

Table 3.8 summarizes heat transfer effectiveness, dimensionless thermal equivalent resistance ($R^+ = R/(mc)^{-1}$) and number of transfer units ($NTU = 1/R^+$) for all the Case I numerical experiments. These two related latter parameters depend on spheres arrangements and HTF mass flow rates. They are all the same for different inlet temperatures. Heat transfer effectiveness for all the arrangements is decreased at higher mass flow rates as a result of lower dimensionless thermal resistivity decrease. Table 3.9 summarizes heat transfer effectiveness, dimensionless thermal equivalent resistance, exergy efficiency, and number of transfer units for three different configurations at a single mass flow rate (0.0052 kg/s) at single inlet (263.15 K) and ambient temperatures (298.15 K). Tables 3.10 and 3.11 summarize heat transfer effectiveness, dimensionless thermal equivalent resistance, exergy efficiency, and number of transfer units for three different

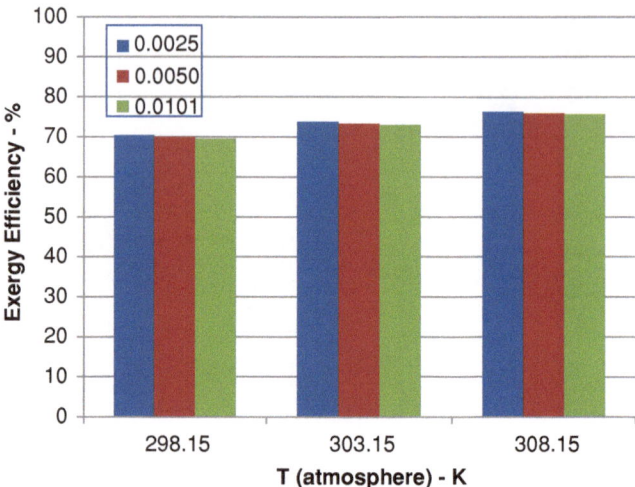

Fig. 3.7 Exergy efficiency at different ambient temperatures and flow rates

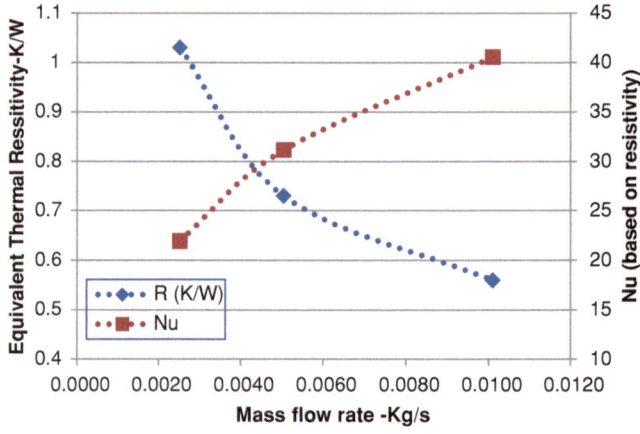

Fig. 3.8 Equivalent thermal resistivity and *Nu* number for all Case I numerical simulations at different flow rates

configurations at two different inlet boundary conditions. As the inlet cross section areas of different arrangement cells are different, all the cases are under different inlet mass flow rates. Heat transfer effectiveness for all the cases relates to the dimensionless thermal resistance. Table 3.12 summarizes equivalent temperature difference, dimensionless thermal resistance, and NTU for SIC capsule arrangement with different diameters.

Table 3.13 summarizes synergy and synergy densities (per unit volume) for different configurations at different inlet conditions. Synergy density can be used

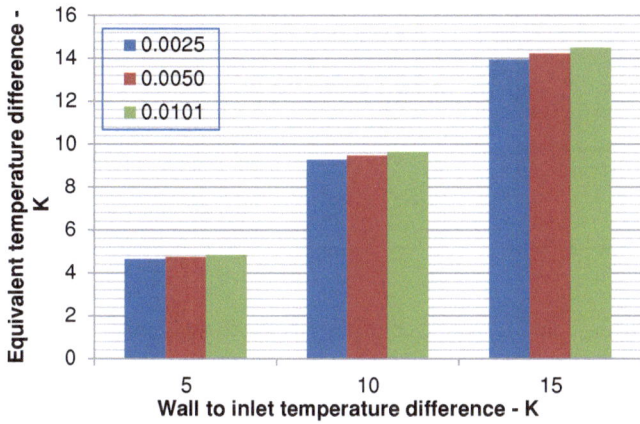

Fig. 3.9 Equivalent temperature difference for all Case I numerical simulations at wall to HTF inlet temperature differences

Table 3.8 Nondimensional thermal resistance and NTU for different mass flow (Case I)

Configuration	For all T_{in}(K): 268.15, 263.15, 258.15		
	Simple cube		
Mass flow rates (Kg/s)	0.0025	0.0050	0.0100
Heat transfer effectiveness	0.14	0.10	0.07
R$^+$	6.52	4.59	3.53
NTU	0.15	0.22	0.28

Table 3.9 Equivalent thermal resistance and temperature difference, nondimensional thermal resistance and NTU at a single mass flow rate (0.0052 kg/s) and inlet temperature (263.15 K) at single ambient temperature (298.15 K)

	R $\left(\frac{K}{W}\right)$	ΔT_{eqv}(K)	R$^+$	NTU	ψ % Exergy efficiency	φ Heat transfer effectiveness
Simple cube	0.75	9.51	9.75	0.10	70	0.10

Table 3.10 Equivalent thermal resistance and temperature difference, nondimensional thermal resistance and NTU (U_{in} = 0.001 m/s, T_{in} = 263.15 K, T_0 = 298.15 K)

	$\dot{m}\left(\frac{Kg}{s}\right)$	R $\left(\frac{K}{W}\right)$	R$^+$	ΔT_{eqv}(K)	ψ % Exergy efficiency	φ Heat transfer effectiveness
Simple cube	0.0050	0.239	2.99	8.58	73	0.29

Table 3.11 Equivalent thermal resistance and temperature difference, nondimensional thermal resistance and NTU (U_{in} = 0.002 m/s, T_{in} = 258.15 K, T_0 = 298.15 K)

	$\dot{m}\left(\frac{Kg}{s}\right)$	R $\left(\frac{K}{W}\right)$	R$^+$	ΔT_{eqv}(K)	ψ % Exergy efficiency	φ Heat transfer effectiveness
Simple cube	0.0100	0.159	3.98	13.34	62	0.22

Table 3.12 Equivalent thermal resistance and temperature difference, nondimensional thermal resistance and NTU with three different diameters ($T_{in} = -10$, $U_{in} = 0.001$)

Capsule diameter—d (m)	$\dot{m}\left(\frac{Kg}{s}\right)$	R $\left(\frac{K}{W}\right)$	ΔT_{eqv}(K)	ψ % Exergy efficiency	φ Heat transfer effectiveness
0.04	0.0013	2.16	9.32	70	0.14
0.08	0.0050	0.73	9.48	70	0.10
0.12	0.0114	0.41	9.59	70	0.08

Table 3.13 Synergy and synergy density for different configurations at different HTF inlet temperatures and mass flow rates

T_{in}(K)	$\dot{m}\left(\frac{Kg}{s}\right) = 0.0026$	
	Synergy $\left(\frac{K}{s}\right) \times 10^6$	Synergy density $\left(\frac{K}{m^3 s}\right) \times 10^3$
268.15	2.19	1.7
263.15	4.38	3.5
258.15	6.56	5.2

as a tool to assess the overall heat transfer performance of the configurations under different mass flow rates and inlet temperatures with different volumes. A lower inlet temperature creates more temperature gradients in the fluid field and corresponds to higher values of synergy density. In cases with higher mass flow rates, the velocity magnitudes will be increases in all directions and as a result leads to higher synergy densities.

3.5.5 Test Cases and the Results: Group II

In this series of tests as Group II, all the domains have a constant wall temperature surface area equal to six complete spherical capsules with a diameter of 0.08 m. The geometry has an extended outlet length equal to two times of a cross section side wall size. All of the runs have two different inlet boundary conditions for momentum equation. These two are constant mass flow rate (= 0.003 kg/s) as Case IV and constant inlet velocity (= 0.0006 m/s) as Case V at constant inlet temperature (263.15 K).

Figure 3.10 shows the temperature values at middle symmetric plane and Fig. 3.11 shows the velocity magnitude on the plane. Different mass flow rates creates different velocity and temperature fields and in the next step a heat transfer analysis based on entransy and synergy of velocity and temperature gradient fields will be done.

Figure 3.12 shows the synergy values at middle symmetric plane for each test case. The maximum synergy strength value areas (red areas) show a favorable coordination between the velocity and temperature gradient fields and it has a

Fig. 3.10 Temperature
values on the middle
symmetric plane

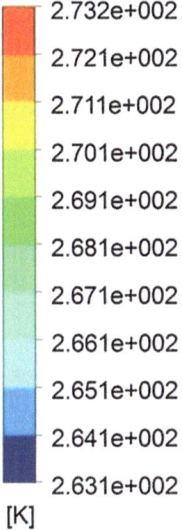
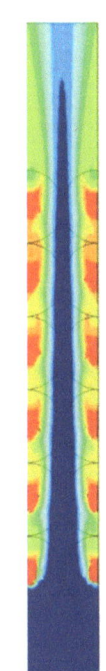

2.732e+002

2.721e+002

2.711e+002

2.701e+002

2.691e+002

2.681e+002

2.671e+002

2.661e+002

2.651e+002

2.641e+002

2.631e+002

[K]

Fig. 3.11 Velocity
magnitudes on the middle
symmetric plane

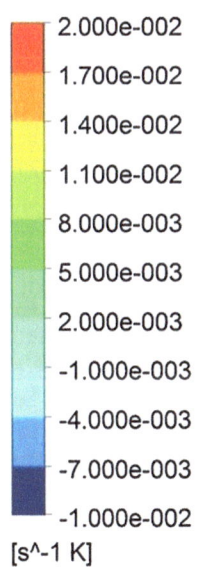
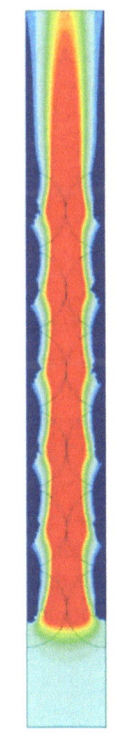

2.000e-002

1.700e-002

1.400e-002

1.100e-002

8.000e-003

5.000e-003

2.000e-003

-1.000e-003

-4.000e-003

-7.000e-003

-1.000e-002

[s^-1 K]

Fig. 3.12 Middle symmetric
plane synergy magnitude

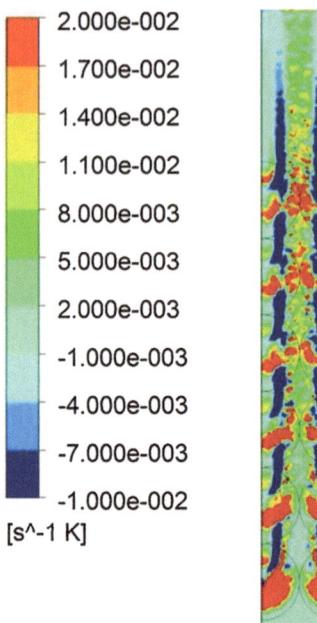

2.000e-002

1.700e-002

1.400e-002

1.100e-002

8.000e-003

5.000e-003

2.000e-003

-1.000e-003

-4.000e-003

-7.000e-003

-1.000e-002

[s^-1 K]

Table 3.14 Synergy and
synergy density (Case V)

$T_{\text{in}}(\text{K})$	$U_{\text{in}}\left(\frac{\text{m}}{\text{s}}\right) = 0.0006$
	$Synergy \left(\frac{\text{K}}{\text{s}}\right) \times 10^5$
	1.35
263.15	$Synergy\ density \left(\frac{\text{K}}{\text{m}^3\text{s}}\right) \times 10^3$
	4.5

different distribution in different geometries and arrangement. Table 3.14 summarize synergy and synergy densities (per unit volume) at different inlet conditions. Synergy density can be used as a tool to assess the overall heat transfer performance of the configuration under the same inlet velocity and temperature.

Table 3.15 summarizes dimensionless thermal resistance, heat transfer effectiveness, exergy efficiency, and NTU for three different capsule arrangements under constant inlet mass flow rate (Case VI). Table 3.16 summarizes dimensionless thermal resistance, heat transfer effectiveness, exergy efficiency, and NTU for three different capsule arrangements under constant inlet velocity (Case V). In both cases a lower dimensionless equivalent thermal resistivity corresponds to higher heat transfer effectiveness value.

Table 3.15 Nondimensional thermal resistance, heat transfer effectiveness, exergy efficiency, and NTU for three different capsule arrangements under similar inlet boundary conditions (Case VI)

Configuration	T_{in}(K): 263.15 SIC
Mass flow rates (Kg/s)	0.003
$R\left(\frac{K}{W}\right)$	0.16
ΔT_{eqv}(**K**)	8.04
R^+	2.05
NTU	0.49
Heat transfer effectiveness—φ	0.40
Exergy efficiency—ψ %	73

Table 3.16 Nondimensional thermal resistance and NTU for three different capsule arrangements with their corresponding first and second law efficiencies under similar inlet boundary condition (Case V)

	$R\left(\frac{K}{W}\right)$	ΔT_{eqv}(K)	R^+	NTU	ψ % Exergy efficiency	φ Heat transfer effectiveness
SIC	0.16	8.04	2.05	0.49	73	0.40

3.6 Concluding Remarks

There are many possible capsule and tank geometry types and arrangements. Any TES tank can be a heat exchanger and before any exergetic analysis on the whole system and TES integration to a bigger system arrangements, e.g., A/C units with its components, it is very helpful to assess the performance of the capsules arrangements, shapes and flow field configurations. In many cases, mass flow rates and inlet temperature of HTF and also PCM phase change temperature is limited to available options and current technology. So, a tool will be needed to assess the heat transfer that takes place between the capsules and the HTF and how efficient the process is. In the next step, any shape and configuration change contributions can be assessed and help the designers and manufacturers to offer more energy saving TES systems.

Thermal assessment parameters for TES system performance have been introduced including nondimensional TES equivalent thermal resistance, equivalent temperature (potential) difference, number of transfer unit (NTU) and energy density through a heat exchanger approach.

References

1. Bejan A (1996) Entropy generation minimization. CRC Press Inc., New York
2. Dicer I, Rosen MA (2002) Thermal energy storage. Wiley, London
3. Erek A, Dincer I (2009) A new approach to energy and exergy analyses of latent heat storage unit. Heat Trans Eng 30:506–515

4. Morgan M, Shapiro H (2008) Fundamentals of engineering thermodynamics. Wiley, New York
5. Guo Z, Li D, Wang B (1998) A novel concept for convective heat transfer enhancement. Int J Heat Mass Transfer 41:2221–2225
6. Guo Z, Tao W, Shah R (2005) The field synergy (coordination) principle and its applications in enhancing single phase convective heat transfer. Int J Heat Mass Transfer 48:1797–1807
7. Tao W, Guo Z, Wang B (2002) Field synergy principle for enhancing convective heat transfer—its extension and numerical verifications. Int J Heat Mass Transfer 45:3849–3856

Chapter 4
Application of Entransy Theory in Absorption Refrigeration System

Abstract In this chapter, energy and entransy analyses are applied to a single stage absorption refrigeration system, the entransy dissipation rate and entransy-dissipation based thermal resistance rate of each component, as well as the system performance on a series of operating conditions, are calculated. The results of the entransy dissipation rate and corresponding thermal resistance analysis can be applied as a useful tool for evaluation and improvement of the absorption refrigeration system. It provides a simple and effective method to identify how to decrease the thermal resistance and increase the heat transfer efficiency at different devices and where a given design should be modified for the best performance.

Keywords Absorption refrigeration system · Coefficient of performance (COP) · Ammonia · Entransy · Heat transfer · Desorption

In recent years, thermal driven absorption refrigeration systems have become an accepted and sound alternative for cooling residential, light commercial and industrial applications. Absorption cooling systems provide great opportunities for energy saving, unlike conventional compression refrigeration system, absorption chiller can use low grade heat energy instead of electricity to produce cooling. This feature makes it possible to use non-conventional sources of energy such as waster heat, geothermal or solar energy as the primary energy input. Furthermore, the working solution pairs used by normal absorption refrigeration systems are environmentally friendly and do not deplete the ozone layer of the atmosphere. All these contributing factors encouraged engineers and researchers to consider absorption systems as important alternatives to vapour compression refrigeration.

The most common absorption systems use NH_3–H_2O or H_2O–LiBr as absorbent-refrigerant pair. Absorption machines based on H_2O–LiBr are typically configured as water chillers for air-conditioning systems because this working fluid utilizes water as the refrigerant, while the NH_3–H_2O cycle is often used as the refrigeration system because of its low evaporation temperature of ammonia. However, both working fluid mixtures had some disadvantages to limit their

J. Gu and Z. Gan, *Entransy in Phase-Change Systems*,
SpringerBriefs in Thermal Engineering and Applied Science,
DOI: 10.1007/978-3-319-07428-3_4, © The Author(s) 2014

applications. The disadvantage of the H₂O–LiBr system came from its negative pressure and corrosion, whereas the main disadvantage of the NH₃–H₂O system was its high water content in the vapor phase ammonia which requires an auxiliary rectifier to separate it. Furthermore, the NH₃–H₂O system exhibits a relatively low coefficient of performance (COP).

In this case, some new absorption system used salt mixture like Ammonia-Sodium Thiocyanate or Ammonia-Lithium Nitrate as the absorbent solution pair has been developed for ice-makers in fishery industry or other similar area. The major active component of the absorbent for these kinds of system is salt. Ammonia (NH₃) is chosen as refrigerant. In this chapter, an NH₃–NaSCN absorption refrigeration system will be studied and a sample performance calculation of this system will be discussed, using entransy concept. Among the first studies on an NH₃–NaSCN system are those of Blytas and Daniels [1] and Sargent and Beckman [2]. Tyagi [3] and Aggarawal et al. [4] provided a review on ammonia-salts absorption refrigeration systems. Thermodynamic and physical properties for the NH₃–NaSCN solution have been presented by Infante [5]. Theoretical and experimental analysis of the NH₃–NaSCN absorption system has been provided by Zhu et al. [6], while interesting energy and energy approach of NH₃–NaSCN heating and cooling system may be found in the works of Zhu and Gu [7].

4.1 Thermodynamic Analysis of NH₃–NaSCN Absorption System

In this section, energy and mass balances are applied to each component of a single stage NH₃–NaSCN absorption system and combined with the state equations for the thermodynamic properties of the working fluids to provide with a set of equations to describe the system characteristics. In order to determine the heat and mass transfer properties of the system, each component is taken as a single unit, the balance equations of mass, energy and entransy dissipation equation for each components include the generator, condenser, evaporator, absorber and Generator & Absorber Heat Exchanger (GAX) were calculated separately.

4.1.1 Fluid Properties

The performance and efficiency of the absorption system are determined to a large degree by the properties of the working fluids. In this analysis presented here, ammonia is the refrigerant, and sodium thiocyanate is the absorbent, therefore the thermodynamic properties for ammonia and ammonia/sodium thiocyanate solution were studied. The properties of ammonia-sodium thiocyanate solution have been studied by Infante [5], while the properties of saturated and superheated vapor and liquid ammonia properties are provided by Sun [8] and Zhu et al. [6].

The relation among saturation equilibrium pressure, concentration and temperature is given as:

$$\ln P = A + \frac{B}{T} \tag{4.1}$$

where

$$A = 15.7266 - 0.298628 \cdot X, \tag{4.2}$$

$$B = -2548.65 - 2621.92 \cdot (1 - X)^3. \tag{4.3}$$

The relation among temperature, concentration and enthalpy is,

$$h = A + B(T - 273.15) + C(T - 273.15)^2 + D(T - 273.15)^3 \tag{4.4}$$

where

$$A = 79.72 - 1072 \cdot X + 1287.9 \cdot X^2 - 295.67 \cdot X^3, \tag{4.5}$$

$$B = 2.4081 - 2.2814 \cdot X + 7.9291 \cdot X^2 - 3.5137 \cdot X^3 \tag{4.6}$$

$$C = 10^{-2} \left(1.255 \cdot X - 4 \cdot X^2 + 3.06 \cdot X^3 \right), \tag{4.7}$$

$$D = 10^5 \left(-3.33 \cdot X + 10 \cdot X^2 - 3.33 \cdot X^3 \right). \tag{4.8}$$

The specific heat data for solutions of sodium thiocyanate in liquid ammonia was presented by Blytas and Daniel and Sargent and Beckman, and was correlated together with the pure liquid ammonia data by Infante [5] to give:

$$c_p = A + B(T - 273.15) + C(T - 273.15)^2, \tag{4.9}$$

where

$$A = 0.24081 - 0.22814 \cdot X + 0.79291 \cdot X^2 - 0.35137 \cdot X^3, \tag{4.10}$$

$$B = 10^{-1} \left(0.251 \cdot X - 0.8 \cdot X^2 + 0.612 \cdot X^3 \right), \tag{4.11}$$

$$C = 10^{-3} \left(-0.1 \cdot X + 0.3 \cdot X^2 - 0.1 \cdot X^3 \right). \tag{4.12}$$

Within the ranges of pressure and temperature concerning refrigeration applications, the two phase equilibrium pressure and temperature of the refrigerant ammonia are linked by the relation,

$$P = 10^3 \sum_{i=0}^{6} a_i (T - 273.15)^i \qquad (4.13)$$

The specific enthalpies of saturated liquid and vapor ammonia are expressed in terms of temperature,

$$h_1 = \sum_{i=0}^{6} b_i (T - 273.15)^i \qquad (4.14)$$

$$h_v = \sum_{i=0}^{6} c_i (T - 273.15)^i \qquad (4.15)$$

The above three equations are fitted by Sun [8] with source data taken from ASHRAE handbook [9]. The coefficients a_i, b_i and c_i are listed in Table 4.1.

The relation between temperature and specific enthalpies of superheated vapor ammonia is,

$$h_{\text{superheated}}(T_2) = h_{\text{saturated}}(T_1) + \int_{T_1}^{T_2} c_p dT, \qquad (4.16)$$

Where T_1 is the saturated temperature corresponding to the pressure of this superheated vapor ammonia, $h_{\text{saturated}}(T_1)$ is the corresponding enthalpy, T_2 is temperature of the super-heated vapor ammonia, $h_{\text{superheated}}(T_2)$ is the corresponding enthalpy. and c_p is the specific heat,

$$c_p = \sum_{i=0}^{6} d_i (T - 273.15)^i \qquad (4.17)$$

The above equations are collected by Zhu et al. [6] with source data taken from the properties of R-717 (ANHYDROUS AMMONIA) handbook presented by Industrial Refrigeration Consortium at University of Wisconsin [10], and the coefficient d_i is listed in Table 4.1.

4.1.2 System Description and Analysis

Figure 4.1 illustrates the main components of the system. High-pressure liquid refrigerant (2) from the condenser passes into the evaporator through an expansion valve V2 that reduces the pressure of the refrigerant to the low pressure level as in the evaporator. The liquid ammonia vaporizes in the evaporator by absorbing heat from the material being cooled and resulting low pressure vapor ammonia (4)

Table 4.1 Coefficients of Eqs. 4.13, 4.14, 4.15 and 4.17

i	a_i Eq. 4.13	b_i Eq. 4.14	c_i Eq. 4.15	d_i Eq. 4.17
0	4.2871×10^{-1}	1.9879×10^{2}	1.4633×10^{3}	1.7467×10^{4}
1	1.6001×10^{-2}	4.4644×10^{0}	1.2839×10^{0}	-3.3129×10^{2}
2	2.3652×10^{-4}	6.2790×10^{-3}	-1.1501×10^{-2}	2.6189×10^{0}
3	1.6132×10^{-6}	1.4591×10^{-4}	-2.1523×10^{-4}	-1.1045×10^{-2}
4	2.4303×10^{-9}	-1.5262×10^{-6}	1.9055×10^{-6}	2.6214×10^{-5}
5	-1.2494×10^{-11}	-1.8069×10^{-8}	2.5608×10^{-8}	-3.3202×10^{-8}
6	1.2741×10^{-13}	1.9054×10^{-10}	-2.5964×10^{-10}	1.7539×10^{-11}

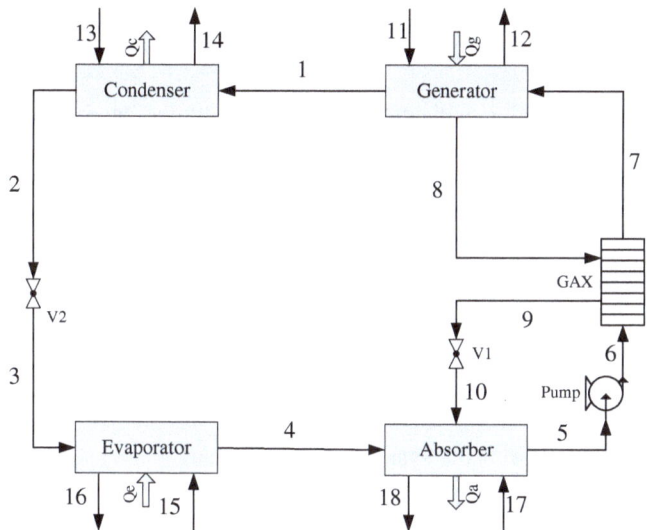

Fig. 4.1 The schematic of a single stage NH₃–NaSCN absorption cycle

passes to the absorber, where it is absorbed by the strong solution (10) coming from the generator through the generator absorber heat exchanger (GAX), and form the weak solution (5). The weak solution is pumped to the GAX and then the generator, and the solution is boiled in the generator. The remaining solution (8) flows back to the absorber and the superheated vapor ammonia (1) passes into the condenser and liquefied to high pressure liquid ammonia by releasing heat to the cooling material and, thus, completes the cycle. Where stream (11) and (12) express the inlet and outlet of the input heat source, Stream (13) (14) and (17) (18) are the corresponding input and output of cooling water supplied to the condenser and absorber (15) and (16) is the input and output of chilled water which will supply cooling to the surrounding environment. The function of GAX is to improve system performance.

Based on the properties of the working solution pair, the equations of mass and energy conservation are determined to describe the heat and mass transfer in every component. The governing mass conservation equations for a steady state are,

$$\sum \dot{m}_i - \sum \dot{m}_o = 0, \tag{4.18}$$

$$\sum (\dot{m}_i X_i) - \sum (\dot{m}_o X_o) = 0. \tag{4.19}$$

The energy balance of each component based on the first law of thermodynamic is,

$$\sum (\dot{m}_i h_i) - \sum (\dot{m}_o h_o) + \left(\sum Q_i - \sum Q_o \right) + W = 0 \tag{4.20}$$

where subscripts i and o indicate corresponding "inlet" and "outlet" of each component, $\dot{m}$ is the mass flow rate, X is the mass concentration of ammonia in the solution, and h is the enthalpy of the working fluid.

The system cooling and heating COP can be written,

$$\text{COP}_{\text{cooling}} = \frac{Q_E}{Q_G} = \frac{\dot{m}_4 (h_4 - h_3)}{\dot{m}_1 h_1 + \dot{m}_8 h_8 - \dot{m}_7 h_7} \tag{4.21}$$

$$\text{COP}_{\text{heating}} = \frac{Q_A + Q_C}{Q_G} = \frac{\dot{m}_1 (h_1 - h_2) + \dot{m}_4 h_4 + \dot{m}_{10} h_{10} - \dot{m}_5 h_5}{\dot{m}_1 h_1 + \dot{m}_8 h_8 - \dot{m}_7 h_7} \tag{4.22}$$

The circulation ratio of the system can be defined,

$$f = \frac{\dot{m}_7}{\dot{m}_1} \tag{4.23}$$

According to the definition of the entransy and entransy dissipation explained in previous section, the entransy dissipation rate for each major component can be expressed,

$$\dot{G}_G = \frac{1}{2} \dot{m}_7 \text{Cp}_7 T_7^2 - \left(\frac{1}{2} \dot{m}_1 \text{Cp}_1 T_1^2 + \frac{1}{2} \dot{m}_8 \text{Cp}_8 T_8^2 \right) + \frac{1}{2} \dot{m}_{11} \text{Cp}_{11} \left(T_{11}^2 - T_{12}^2 \right), \tag{4.24}$$

$$\dot{G}_C = \frac{1}{2} \dot{m}_1 \text{Cp}_1 T_1^2 - \frac{1}{2} \dot{m}_1 \text{Cp}_{2v} T_2^2 + \frac{1}{2} \dot{m}_1 T_2^2 (\text{Cp}_{2v} - \text{Cp}_{2L})$$
$$+ \frac{1}{2} \dot{m}_{13} \text{Cp}_{13} \left(T_{13}^2 - T_{14}^2 \right), \tag{4.25}$$

$$\dot{G}_E = \frac{1}{2}\dot{m}_3 Cp_3 T_3^2 - \frac{1}{2}\dot{m}_3 Cp_{4V} T_4^2 + \frac{1}{2}\dot{m}_3 T_4^2 (Cp_{4v} - Cp_{4L}) \\ + \frac{1}{2}\dot{m}_{15} Cp_{15}(T_{15}^2 - T_{16}^2),$$

(4.26)

$$\dot{G}_A = \frac{1}{2}\dot{m}_4 Cp_{4V} T_4^2 + \frac{1}{2}\dot{m}_{10} Cp_{10} T_{10}^2 - \frac{1}{2}\dot{m}_5 Cp_5 T_5^2 + \frac{1}{2}\dot{m}_{17} Cp_{17}(T_{17}^2 - T_{18}^2),$$

(4.27)

$$\dot{G}_{GAX} = \frac{1}{2}\dot{m}_8 Cp_8 (T_8^2 - T_9^2) + \frac{1}{2}\dot{m}_6 Cp_6 (T_6^2 - T_7^2).$$

(4.28)

The above equations can be used to calculation the entransy value for sample heat exchange process. However, in current absorption refrigeration cycle, there are phase change in the condenser and evaporator parts, and also energy release in sorption process. To indicate the entransy calculation for phase change process, the entransy value was re-defined,

$$\dot{G} = \frac{1}{2}\dot{m}\, C\, pT^2 = \frac{1}{2}\dot{m}\, hT,$$

(4.29)

where in this equation the value of "CpT" was replace by the enthalpy "h".

And also to simplify the calculation in sorption process, the solution concentration effect was neglected, the whole process was treated like a partial condensation. Only the effect of liquid and vapour ammonia was taken into account.

4.1.3 Performance Simulation of NH$_3$–NaSCN Absorption System

In order to make the simulation model of the NH$_3$-NaSCN absorption system convenient and simplified, a set of basic assumptions are made:

- The system is in steady state and the heat loss or heat gain from the environment is neglected.
- The liquid ammonia leaving the condenser and vapor ammonia leaving the evaporator are both saturated.
- The electricity consumption of the solution pump is negligible.
- The strong solution leaving the generator and the weak solution leaving the absorber are both saturated. The generating temperature and absorbing temperature are treated as the corresponding outlet solution temperature of generator and absorber.
- Flow resistance, pressure losses and heat losses in pipes and components are ignored. There is no heat transfer because of radiation, viscous dissipation, pressure gradients, concentration gradients or chemical reactions.

Table 4.2 The reference point for the absorption system simulation program

Unit	Description	Value
Absorber	Inlet cooling water (Absorber part) temperature	21.7 °C
Condenser	Inlet cooling water (Condenser part) temperature	23 °C
Generator	Inlet hot water temperature	99.5 °C
Evaporator	Outlet chilled liquid temperature	0.2 °C
Solution pump	Mass flow rate of the weak solution	0.07 kg/s
GAX	Effectiveness of the heat exchanger	80 %

Based on the above assumptions and the heat and mass conservation equations, entransy dissipation equations and the state equations for thermodynamic properties, a computer simulation program was built to investigate the mass, energy and entransy analysis of the system in different operating conditions. A design operating condition, which includes the heating, cooling and chilled water inlet temperatures, the mass flow rate of weak solution leaving the solution pump from the absorber and the heat exchanger efficiency of each heat transfer component, was selected to be a reference point for evaluation of the effect of different parameters. Then a wide range of operating conditions was selected and simulated for different components by changing the corresponding parameter while keeping others consistently with the reference point. The simulation results will be discussed in the next section. The reference point is described in Table 4.2.

4.2 Results and Discussion

With the given parameters and assumptions, the simulation program calculates the values of temperature, enthalpy, mass flow rate, solution concentration, entransy dissipation rate and entransy-dissipation based thermal resistance at all points of the absorption cycle. The simulation results for the reference point are listed in Table 4.3.

The calculation results of heat transfer rate, entransy dissipation rate and entransy-dissipation based thermal resistance for each main component are presented in Table 4.4. It is shown that the highest heat transfer rate and entransy dissipation rate occur in the generator and these two values in the condenser is slightly higher than those in the evaporator part, this is primarily due to the superheating property of the inlet vapor ammonia in the condenser. And also the highest entransy-dissipation based thermal resistance rate occurs in the condenser, while the evaporator has the lowest thermal resistance. Variation of the entransy dissipation rate and thermal resistance rate for each component for various operating conditions will be comprehensively examined below.

Figure 4.2 presents that the variation of the entransy dissipation rate in the generator part versus different heating source temperatures. In this case, the hot

Table 4.3 Absorption system data obtained from the thermodynamic analysis

Point	T (°C)	h (kJ/kg)	m (kg/s)	X (% NH₃)	G (kW·K)
1	90	1689.7	0.0088	100	2706.55
2	25	315.88	0.0088	100	415.41
3	25	315.88	0.0088	100	415.41
4	−10	1449.5	0.0088	100	1682.45
5	25	−102.06	0.07	46.22	−1063.88
6	25	−102.06	0.07	46.22	−1063.88
7	67.82	15.53	0.07	46.22	185.14
8	90	71.78	0.0612	38.45	796.39
9	38	−62.79	0.0612	38.45	−596.89
10	38	−62.79	0.0612	38.45	−596.89
11	99.5	417.43	0.67	0	52111.02
12	93.03	390.35	0.67	0	47884.43
13	23	96.96	1	0	14356.38
14	25.87	109.02	1	0	16299.35
15	5	21.62	0.5	0	1503.12
16	0.24	1.58	0.5	0	108.01
17	21.74	91.99	1	0	13531.03
18	24.14	101.77	1	0	15127.67

Table 4.4 Energy transfer, entransy dissipation and thermal resistance rate at the various components

Components	Heat Transfer (kW)	Entransy Dissipation (kW·K)	Thermal Resistance (K/kW)
Generator	18.2	908.79	2.74
Condenser	12.12	348.17	3.48
Evaporator	10	128.07	1.28
Absorber	16.09	552.8	2.14
GAX	8.22	186.34	2.76

water inlet temperature increasing from 72 °C to 100 °C, while keep other input parameters constant. The simulation results shown that with the increase of the heating source inlet temperature, the entransy dissipation rate in the generator also increase linearly. It is reasonable because a higher hot water temperature will course a higher heat transfer rate in the generator part. On the other words, more heat will be absorbed by the inside solution in a higher heating source temperature. As a result, this higher heat flux rate will course a higher entransy dissipation rate.

Figure 4.3 shows the variation of the variation of the entransy-dissipation based thermal resistance rate in the generator with different heating source temperatures. It is shown that the entransy-dissipation based thermal resistance rate will decrease with the increase of the heating source inlet temperature. When the heating source temperature is lower than 75 °C, the thermal resistance rate is pretty big and decrease quickly with the increase of hot water temperature. Once the generator

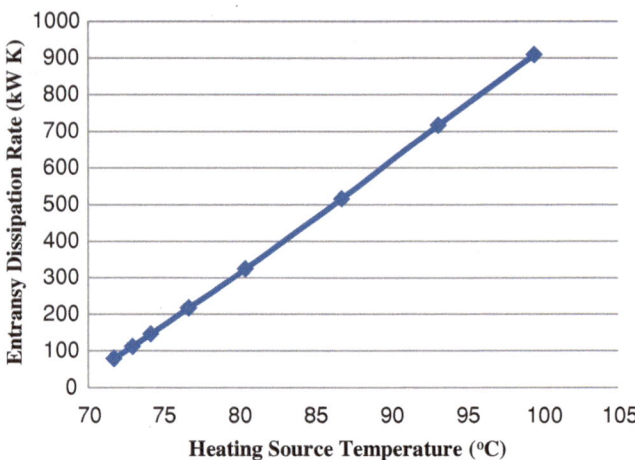

Fig. 4.2 Variation of the entransy dissipation rate in the generator with the heating source temperature

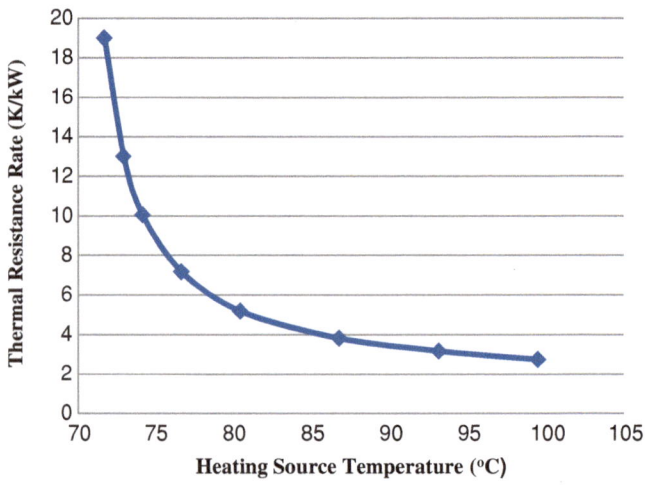

Fig. 4.3 Variation of the entransy-dissipation based thermal resistance rate in the generator with the heating source temperature

having a hot water inlet temperature higher than 75 °C, the thermal resistance rate will increase with the hot water temperature in a very slow pace.

Figures 4.4 and 4.5 indicate the variation of the entransy dissipation rate and entransy-dissipation based thermal resistance rate in the condenser part with the changers of the condenser cooling water inlet temperature. In this case we keep other components' parameters constant, only increase the condenser cooling water inlet temperature from 7.5 to 41.5 °C with constant flow rate. The results show that

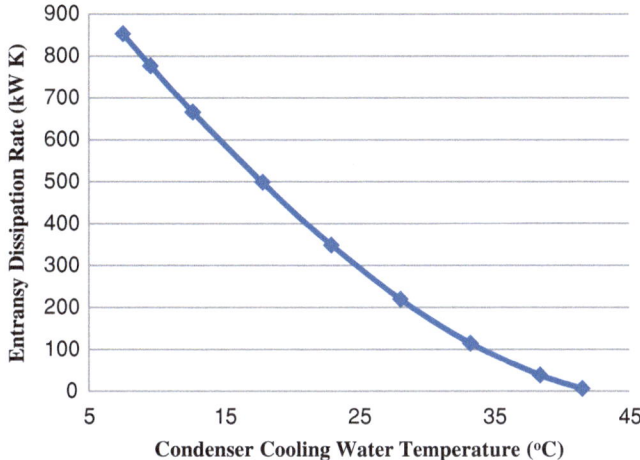

Fig. 4.4 Variation of the entransy dissipation rate in the condenser with the condenser cooling water temperature

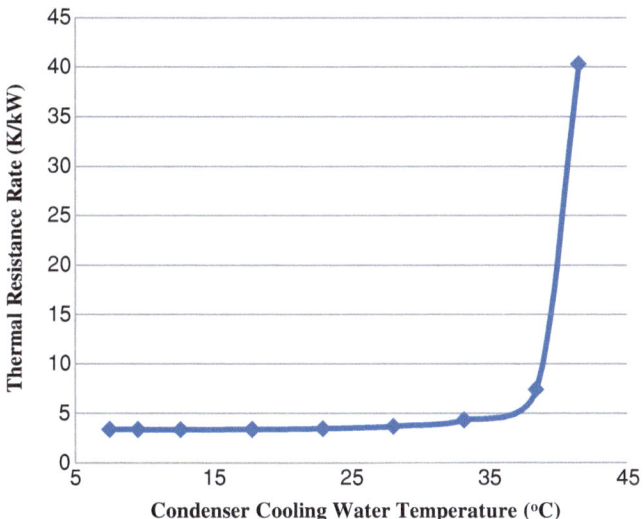

Fig. 4.5 Variation of the entransy-dissipation based thermal resistance rate in the condenser with the condenser cooling water temperature

with the increasing of the cooling water temperature, the entransy dissipation rate of the condenser decreases from 850 kWK to a value smaller than 10 kWK. On the other hand, the entransy-dissipation based thermal resistance rate in the condenser part increase with the increasing of the cooling water temperature. Once the cooling water inlet temperature lower than 35 °C, the change of this value is very small, if the condenser cooling water inlet temperature increase above 35 °C, the

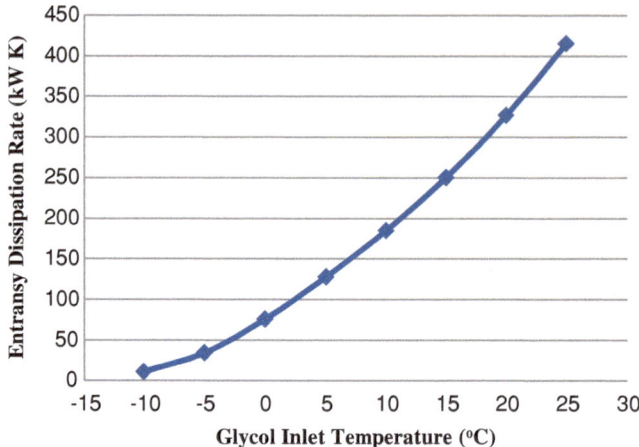

Fig. 4.6 Variation of the entransy dissipation rate in the evaporator with the glycol liquid inlet temperature

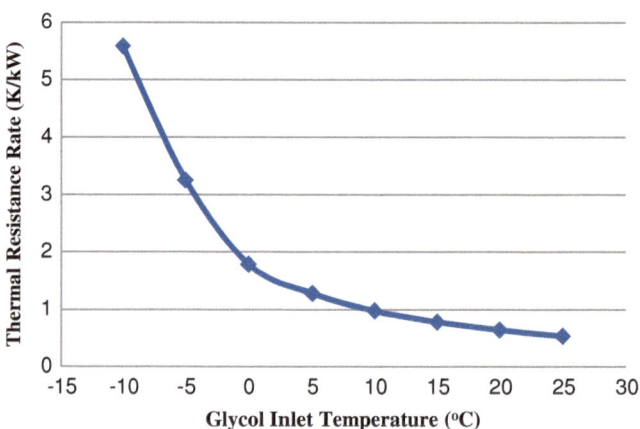

Fig. 4.7 Variation of the entransy-dissipation based thermal resistance rate in the evaporator with the glycol liquid temperature

thermal resistance rate will increase dramatically, it will makes this system highly impractical to be operated at this condition. This result also straight follow the COP calculation based on the first law of thermodynamic. Once the condenser cooling water temperature is higher than a value, 35 °C in this case, the system COP will decrease to an unacceptable value to indicate that the system cannot be operated in this status.

Figures 4.6 and 4.7 explain the variation of the entransy dissipation rate and entransy-dissipation based thermal resistance rate in the evaporator with the change of the glycol inlet temperature. In this simulation, we changer the

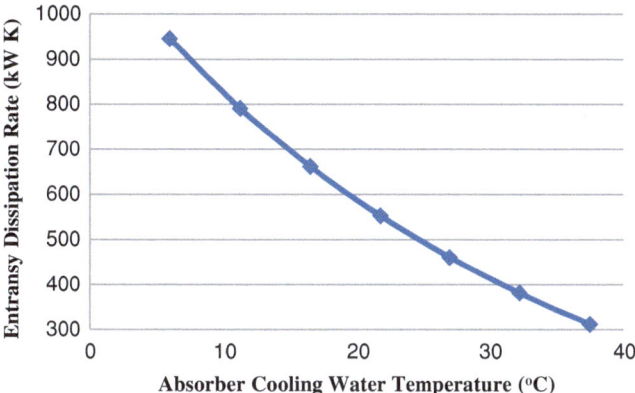

Fig. 4.8 Variation of the entransy dissipation rate in the absorber with the absorber cooling water inlet temperature

evaporator temperature from −23°C to 10°C, keep a 15°C difference between the glycol inlet temperature and the evaporator temperature. And keep other parameters constant including the flow rate of the glycol liquid. The results shown that, with the increase of the glycol inlet temperature, corresponding to the increase of the evaporator temperature, the entransy dissipation rate of the evaporator part will increase from 10 to 415 kWK, the reason of the change is, with the increase of the evaporator, the system cooling capacity will increase also. It means more heat will be exchanged at a higher evaporator temperature; as a result, it leads to a higher entransy dissipation value.

From the indication of Fig. 4.7, the entransy-dissipation based thermal resistance rate will decrease with the increasing of the glycol inlet temperature. It means the heat is easier to be transferred in a higher evaporator temperature. It also can be approved by the change of the system COP and cooling capacity changes. At the high evaporator temperature, the thermal resistance is small, it means the energy is easy to transfer from one side of the heat exchanger to another. As a result, both the cooling capacity and system COP can be maintained in a higher level.

The variation of the entransy dissipation rate and entransy-dissipation based thermal resistance rate in the absorber with the change of the absorber cooling water temperature was shown in Figs. 4.8 and 4.9. With the increasing of the absorber cooling water temperature, the entransy dissipation rate of the absorber part will decrease linearly, while the entransy-dissipation based thermal resistance rate will increase. If the cooling water inlet temperature is higher than 35 °C in this case, the thermal resistance rate will increase sharply and the heat transfer efficiency of the absorber part will decrease quickly, this will greatly influence the system performance and even make current system unacceptable to be operated in some conditions.

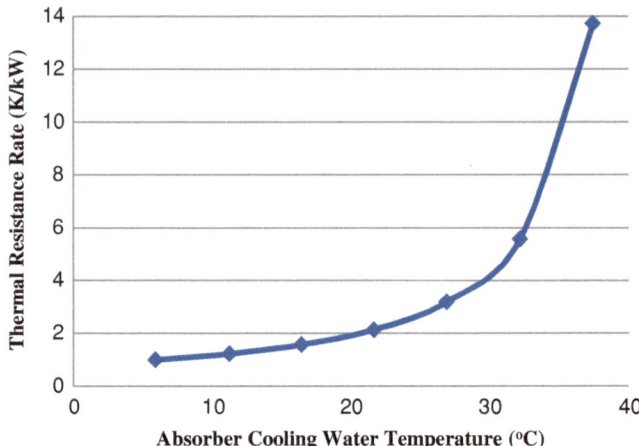

Fig. 4.9 Variation of the entransy-dissipation based thermal resistance rate in the absorber with the absorber cooling water inlet temperature

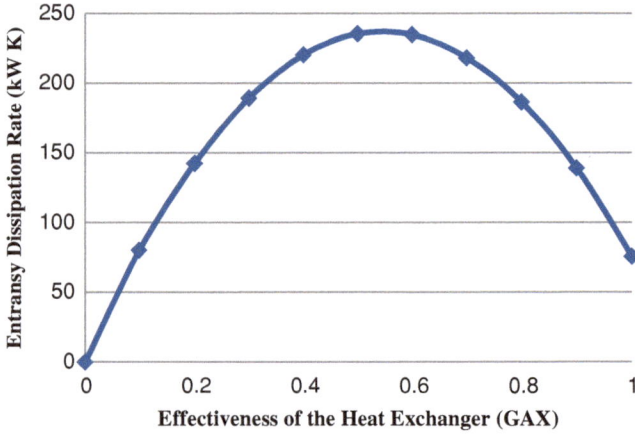

Fig. 4.10 Variation of the entransy dissipation rate in the GAX with the effectiveness of the heat exchanger

Figure 4.10 indicates the variation of the entransy dissipation rate in the GAX with the changes of the heat exchanger effectiveness. In other components, the trend of the entransy dissipation rate always increase or decrease with the constant change of the related inlet temperature. In the GAX part, the value of the entransy dissipation rate will first increase together with the increasing of the heat exchanger effectiveness, achieve a peak value when the heat transfer efficiency reach a value about 0.55, then change the trend to decrease with the increase of the heat exchanger effectiveness.

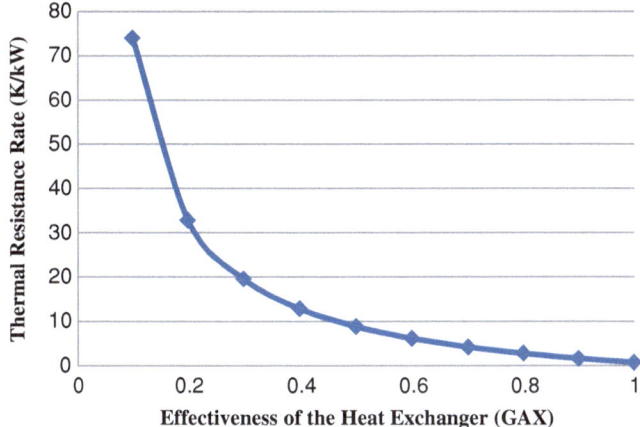

Fig. 4.11 Variation of the entransy-dissipation based thermal resistance rate in the GAX with the effectiveness of the heat exchanger

Figure 4.11 shows the variation of the entransy-dissipation based thermal resistance rate in the GAX part with the change of the effectiveness of the heat exchanger. It can be seen that although the trend of the entransy dissipation rate shown in Fig. 4.10 like a parabola, the value change of the thermal resistance is still follow the same character like other components. This curve indicate that the entransy-dissipation based thermal resistance of the GAX will decrease with the increase of the heat exchanger effectiveness. Once the heat transfer efficiency is 0, the value of thermal resistance at this point will be infinite. Otherwise, if the heat transfer efficiency of the GAX is 1, the value of thermal resistance will pretty close to 0.

4.3   Concluding Remarks

In this chapter, the energy and entransy analysis are applied to a single stage NH_3-NaSCN absorption refrigeration system, the entransy dissipation rate and entransy-dissipation based thermal resistance rate of each component as well as the system performance on a series of operating conditions are predicted. Results shown that the entransy dissipation rate and the thermal resistance of each component will change together with the change of corresponding inlet temperature. While the system COP, cooling capacity and heating load will always strictly follow the change of the entransy-dissipation based thermal resistance. The reason is that, with the decrease of the entransy-dissipation based thermal resistance, the heat is easier to be transferred in the related heat exchanger, as a heat exchanger based system, this will greatly increase the heating or cooling capacity, and finally increase the system COP.

The results of the entransy dissipation rate and corresponding thermal resistance analysis presented in this paper can be applied as a useful tool for evaluation and improvement of the NH_3–NaSCN absorption refrigeration system. It provides a simple and effective method to identify how to decrease the thermal resistance and increase the heat transfer efficiency at different devices and where a given design should be modified for the best performance.

References

1. Blytas G, Daniels F (1962) Concentrated solutions of NaSCN in liquid ammonia: solubility, density, vapor pressure, viscosity, thermal conductance, heat of solution, and heat capacity. J Am Chem Soc 84:1075–1083
2. Sargent S, Beckman W (1968) Theoretical performance of an ammonia–sodium thiocyanate intermittent absorption refrigeration cycle. Sol Energy 12:137–146
3. Tyagi K (1984) Ammonia-salts vapour absorption refrigeration systems. Heat Recovery Syst CHP 4:427–431
4. Aggarawal M, Aggarawal R, Sastry YVSR (1985) Solid absorbents for solar-powered refrigeration systems. Energy 34:423–426
5. Infante CA (1984) Thermodynamic and physical property data equations for ammonia–lithium nitrate and ammonia–sodium thiocyanate solutions. Sol Energy 32:231–236
6. Zhu L, Wang S, Gu J (2008) Performance investigation of a thermal-driven refrigeration system. Int J Energy Res 32:939–949
7. Zhu L, Gu J (2010) Second law-based thermodynamic analysis of ammonia/sodium thiocyanate absorption system. Renew Energy 35:1940–1946
8. Sun D (1998) Comparison of the performances of NH_3–H_2O; NH_3–$LiNO_3$ and NH_3–NaSCN absorption refrigeration systems. Energy Convers Manag 39:357–368
9. ASHRAE Handbook (1993) Fundamentals, Chap. 17. ASHRAE, Atlanta
10. Properties of R-717 (anhydrous ammonia). Industrial Refrigeration Consortium, University of Wisconsin, Madison